CHARACTERIZATION OF ORGANIC COMPOUNDS BY CHEMICAL METHODS

AN INTRODUCTORY LABORATORY TEXTBOOK

TERENCE C. OWEN

UNIVERSITY OF SOUTH FLORIDA
TAMPA, FLORIDA

1969

MARCEL DEKKER, INC. New York

PREFACE

There is no better device than classical qualitative organic analysis for developing a real understanding and appreciation of organic chemistry. Even so, this part of the curriculum often is deferred until the senior year and even deleted entirely from the undergraduate program. Postponement or elimination is justified primarily on two counts: that organic qual. is "difficult" and requires considerable prior lecture material and bookwork; that no practicing chemist in his right mind would use a classical, noninstrumental approach anyway.

Both arguments are fallacious. True, no chemist would, today, use classical approaches *exclusively*, but optimum effectiveness invariably results from an apt combination of classical and instrumental techniques. While identification of the functionally atypical compounds often selected for advanced work (using any of the several excellent manuals currently available) does require considerable background, the student can cope with simple compounds which behave themselves before he has even completed his study of the paraffins—as long as he is prepared, and encouraged to read ahead. Finally, it would be ridiculous to pretend that the purpose of the beginning undergraduate organic laboratory curriculum is to provide professional practitioners of particular skills—be they preparative, interpretive, or observational—actually developed. Most students will never again utilize these skills. The purpose,

iii

then, is to provide them with the widest range of experience possible and to inculcate an appreciation of the largest number of substances, reactions, and techniques in the shortest possible time and with the minimum expenditure of labor and material. To this end, a properly designed introductory program of qualitative analysis, taken *alongside* early lectures, is by far the best possible approach.

At the University of South Florida, the first year of organic chemistry is composed of three quarters. In the first, the student is introduced to techniques for preparation, isolation, and purification, using specified procedures and a fairly rigid schedule. In the second, the student identifies and characterizes rigorously a number (6–12 or more) of straightforward unknown substances, progressing at his own rate. In the third quarter the work varies somewhat, but typically the student carries out more advanced macro and semimicro preparations, devises his own procedures from minimal directions, and characterizes and examines his products using procedures of his own choosing. A further quarter of advanced qualitative analysis is offered at the senior level for chemistry majors. "Difficult" substances are identified, and the use of a full range of spectroscopic (ultraviolet, infrared, and NMR) as well as classical techniques is required.

The essence of this book, then, is simplicity. It deals with the twelve most familiar functional groups only and, indeed, only those substances which exhibit fairly uncomplicated, typical behavior. No apology is offered for this. The whole purpose is to fix the simple functional-group reactions firmly in the student's mind and to develop his faith that these really are characteristic. Anomalous behavior will present itself soon enough in anything but the most elementary study. The scheme developed forms a sound basis upon which more advanced work may develop by addition and modification but without substitution or elimination. The student who gains confidence in himself and the subject during his early work, and develops an ability to think simple problems through, usually evolves his own advanced analytical scheme by confident adaptation from both textbooks and advanced laboratory manuals. He rarely falls into the otherwise tempting trap of rote and cookbook.

Microtechniques are advocated throughout. The advantages thereof go far beyond the obvious economies of materials, equipment, and laboratory space. Savings of time, that most expensive and irreplaceable asset, are dramatic. Efficient, tidy work habits are obligatory and are readily checked. Adaptation, when necessary, to larger-scale preparative work is very easy.

Contrary to the criticism which might be leveled, work of this kind actually offers more preparative experience than does a conventional preparative organic course. Every derivative is a "prep." and every test a reaction; the production and purification of cyclohexanone 2,4-dinitrophenylhydrazone is equally as significant an exercise as that of, say, diazoaminobenzene. The inherent emphasis on pure products rather than high yields breeds the most desirable work habits and attitudes. Each exercise is a unique personal challenge to the individual student and one with which he proceeds at his own rate. The more able student soon forges ahead, but is slowed down, and at the same time stimulated by exercises of steadily increasing difficulty. The average student proceeds more slowly but benefits tremendously by association with his more able neighbor. Even the weakest at least develops some proper conception of chemistry as an experimental and deductive science.

It is quite impossible to acknowledge all those students, teachers, and colleagues who, directly or indirectly, have influenced this book. I am especially grateful to A. I. Vogel, for his classic tome "Practical Organic Chemistry" which has influenced me more than any other chemistry book I have ever read, to Helena Bradbury for innumerable discussions of qualitative analysis as a pedagogic device, to Adora Winans for handling a tricky manuscript with rare urbanity, and, above all, to the late Ernest Arthur Mason for stimulating a lifelong love of chemistry.

TERENCE C. OWEN

CONTENTS

1

INTRODUCTION

"The theory of evolution was greatly objected to because it made men think."

"To collect fumes of sulfur, hold a deacon over a flame in a test tube."

One purpose of qualitative analysis is to identify an
unknown substance as completely as possible. To the practicing chemist this is usually the sole purpose, to be achieved
as speedily and as economically as possible. To you, the
student, however, this is only one purpose among several, and
a relatively minor one at that. Much more important is its
provision of a vehicle by means of which you may gain the
widest possible practical experience of chemistry in the
shortest possible time and with the minimum of space, equipment, and material. Grade-points obviously result from getting
the right answer, fast, but much greater benefit in both practical and theoretical endeavor comes surprisingly quickly with
the skill and experience of thorough, careful laboratory work.
Organic chemistry is an art as well as a science, and cannot
be mastered solely through textbook and lecture.

1. INTRODUCTION

Each unknown substance, then, is a challenge; a personal, private, individual challenge, to be treated as such and to be mastered with all the knowledge, experience and ingenuity you can muster. If you treat it simply as a job of work, mechanical application of "cook book" procedures will probably give the correct answer (at least in elementary work), but you will rob yourself of the experience and first-hand knowledge which only conscientious laboratory work can give.

Ideally, you should devise your own analytical scheme. Unfortunately, those substances which lend themselves to early textbook treatment often are the most difficult to work with, so that about a year of classwork would be necessary before embarking on laboratory work. Then, given large amounts of space, time, material, ability and patience, you could devise your own tests, find out their limitations, make your own mistakes, and slowly become a thoroughly experienced chemist. Few people have time, money, or inclination for this approach. An alternative is to adopt a scheme devised by others. With a suitably designed scheme, you can proceed much more rapidly, start laboratory and classwork simultaneously, try out some of the reactions for yourself, and generally develop an appreciation of organic chemistry as a study of real things rather than of disembodied spirits. In return for these advantages you surrender some autonomy. You will not, at first, understand some of the chemistry of many of the procedures; neither

will you appreciate the reason for the precise order in which
they must be applied. Apparent failures will occur, since no
universally applicable scheme has yet been devised. It is
imperative that you search perpetually for an understanding of
the chemistry underlying all procedures and recipes , to the
absolute limits of your ability and knowledge at any time. In
other words, think! As a reminder, a few "howlers" are repro-
duced as chapter headings throughout this book. "Howlers"
come from many sources; some are simple typing or spelling
mistakes; many more are half-read or half-heard phrases,
thoughtlessly regurgitated in examinations or notebooks. In
all cases the trivial error converts a simple statement into
ludicrous nonsense. The perpetrator of the howler is held up
to ridicule since it is transparently obvious that he gave his
utterance not an iota of thought. Although howlers in laboratory notebooks are rarely amusing, they instantly reveal the
unthinking student.

It would be ridiculous to start this kind of work by ex-
pecting you to identify any complex substance from among the
hundreds of thousands of organic compounds. Such a task would
(and does!) severely extend even an experienced organic chemist.
In any case, this would defeat the object of the exercise which
is not just to get the answer (the instructor already knows it)
but to establish your faith in the simple functional-group
reactions and to implant these firmly in your mind. Initially,

3

1. INTRODUCTION

therefore, compounds from a limited range of well-known functional-group classes will be used. Each compound is carefully selected to behave properly if your work is done and reasoned carefully. There are no traps! The analytical procedure given is geared to this limited scope and, although designed to be capable of extension, will not suffice for more advanced work without suitable supplementation and experience.

An identification is carried out in two stages. The functional group (if any) attached to the hydrocarbon chain is first identified and most carefully confirmed. Then the compound is converted, by means of predictable reactions of this functional group, into other solid compounds ("derivatives") which are identified uniquely by comparing their melting points with those of known compounds containing the same functional group. Each stage proceeds initially by process of elimination. Following Sherlock Holmes' excellent example, one does not set out to identify the villain, but rather considers all possible suspects and eliminates them successively until only one remains. For example, if a sample is solid, all liquids and gases are eliminated from consideration; if insoluble in water, so are all water soluble substances; if it melts at about 162°C, all possibilities listed as melting more than a few degrees away from this value may be rejected. When but a few possibilities remain, specific tests or determinations are selected (or devised) to afford complete identification.

FUNCTIONAL GROUP ORGANIC CHEMISTRY

"A vacuum is a U tube with a flask at one end."

"A person should take a bath once in the summer and not quite so often in the winter time."

"For drowning: Apply artificial respiration until the patient is dead."

Modern introductory textbooks of organic chemistry usually approach the subject from the standpoint of atomic and molecular structure, valence bonding and reaction mechanism. More useful, however, for purposes of characterization and identification in the laboratory, is a classical consideration of the reactions of the functional groups attached to the carbon skeleton. Since this book is intended for students who may have studied organic chemistry for only a few weeks, this chapter gives a brief summary of reactions of particular value for characterization and identification of the functional groups and the preparation of solid derivatives. The structures of the hydrocarbons are considered first, since all organic compounds contain carbon, but thereafter the groups of compounds are considered in order as they arise in the analytical scheme. Material in this chapter will supplement and should be supplemented by a good textbook.

2. FUNCTIONAL GROUP ORGANIC CHEMISTRY

<u>THE NATURE OF ORGANIC CHEMISTRY</u>

Organic chemistry is the study of the compounds of carbon. In stable organic compounds, carbon almost invariably exhibits a covalence of four and forms stable covalent bonds particularly with hydrogen, with the more electronegative elements at the top right-hand side of the periodic table (nitrogen, oxygen, fluorine, sulfur, chlorine, bromine, iodine), <u>and from one carbon atom to another</u>. This latter bonding characteristic, capable of extension to chains of carbon atoms of almost unlimited length, is unique to this element and is responsible for the infinite diversity of known and possible carbon compounds and, hence, for the subject of organic chemistry. The four bonds are directed towards the corners of a regular tetrahedron (Figure 1a) so that a three-dimensional drawing or model

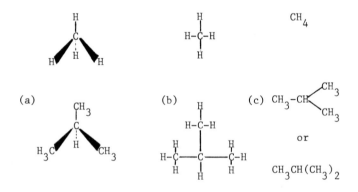

Figure 1

Tetrahedral (a), planar (b), abbreviated (c) structural formulae for methane and isobutane.

is necessary for proper representation of the position of the
atoms in any organic compound. For many purposes, however, and
for all material in this book, the more convenient planar
(Figure 1b) or abbreviated (Figure 1c) structural representa-
tions are adequate.

Covalent bonds are formed by sharing outer-shell electrons
in pairs between the atoms bonded. Thus each radial line join-
ing two atoms (C-H; C=C; C≡C) represents a bonding pair of
electrons. Electronegative elements often have in their outer
shells pairs of electrons which are not involved in bonding.
These "unshared pairs" are represented by concentric bars
(C-$\bar{\underline{O}}$-C; C-$\bar{\underline{F}}$|; C≡N|). A reaction very often involves the sharing
of such an unshared pair with an atom which can accept such a
share, so that a bar in the structural formula becomes a line
($H_3N| + BF_3 \rightarrow H_3\overset{+}{N}-\overset{-}{B}F_3$). Note that the donor atom increases its
positive charge (or decreases its negative charge) by one unit,
while the acceptor does the opposite. Unshared pairs often
are not shown explicitly in structural formulae unless it is
necessary to draw particular attention to them.

THE STRUCTURES OF HYDROCARBONS

Compounds containing only carbon and hydrogen are known as
hydrocarbons. They are divided into two main groups, the aliphatic and the aromatic hydrocarbons, the former being further
subdivided into the saturated and the unsaturated types. The

7

2. FUNCTIONAL GROUP ORGANIC CHEMISTRY

main aromatic subgroup comprises the benzenoid compounds which contain one or more benzene rings to which aliphatic groupings may be attached.

Saturated Aliphatic Hydrocarbons (Alkanes or paraffins and cycloalkanes or cycloparaffins)

All compounds in this group contain only single (two electron) covalent bonds. The alkanes may be regarded as being derived from the simplest hydrocarbon, methane, by re-placement of its hydrogen atoms by carbon atoms each bearing a complement of hydrogen sufficient to saturate its normal tetra-covalence. The structures of the first five such compounds are shown below.

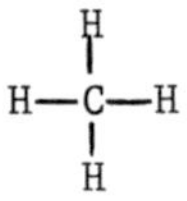

Methane 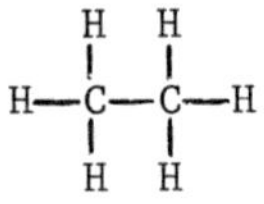   Ethane 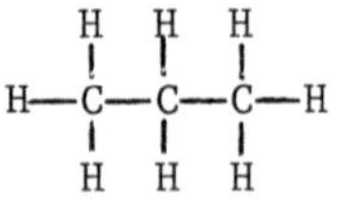   Propane

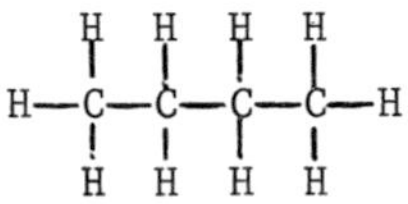

n–Butane 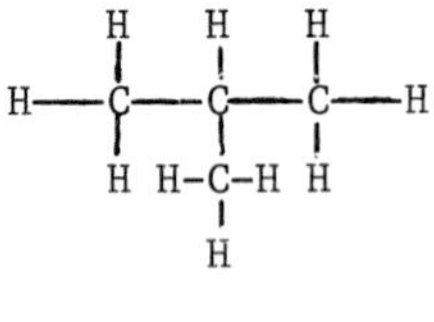   Isobutane

Each alkane has the molecular formula C_nH_{2n+2} where n is an integer. There is only one possible structure for methane, CH_4, one for ethane, C_2H_6, and one for propane, C_3H_8; only one compound corresponding to each molecular formular is known. Two

structures are possible for C_4H_{10}, however, and two distinct
compounds, n-butane and isobutane, exhibiting different physical properties, are known. Three structures are possible for
pentane, C_5H_{12}, five for hexane, C_6H_{14}, nine for heptane,
C_7H_{16}, eighteen for octane, C_8H_{18}, and so on. The number of
structures possible for the higher members of the series
rapidly becomes very large (355 for $C_{12}H_{26}$; 316,319 for $C_{20}H_{42}$)
and each represents a distinct substance. Such compounds,
having the same molecular formula but different structures and
physical properties, are called _isomers_; the phenomenon is
known as _isomerism_. It occurs with all groups of organic compounds so that the total number of such structural possibilities is virtually limitless.

Cycloalkane structures contain one or more rings of carbon
atoms.

Cyclopropane   Cyclobutane   Cyclohexane

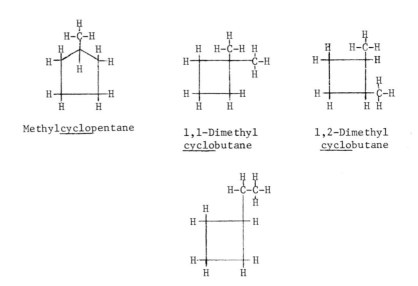

Methylcyclopentane   1,1-Dimethyl   1,2-Dimethyl
                          cyclobutane   cyclobutane

Ethylcyclobutane

Cycloalkanes containing one ring (monocyclic) all have molecular formula C_nH_{2n}, bicyclic ones are C_nH_{2n-2}, and so on. Many
additional possibilities for isomerism arise in this series.
Thus C_6H_{12} represents cyclohexane, methylcyclopentane, ethylcyclobutane, two propylcyclopropanes and the various dimethyl-
cyclobutanes, trimethylcyclopropanes, and methylethylcyclopro-
panes. The carbon-carbon and carbon-hydrogen single bonds are
very stable so that the alkanes and cycloalkanes are not reactive to the common reagents and the usual reaction conditions
used in the laboratory.

Benzene

     The hydrocarbon benzene is unique. It is a monocyclic
compound containing a ring of six carbon atoms each bearing

one hydrogen. Its molecular formula, C_6H_6, suggests a considerable degree of unsaturation such as is implicit in the
Kekulé structure but the substance does not exhibit the
characteristic reactivity of alkenes. It is hydrogenated (to
cyclohexane) only under forcing conditions and resists reaction with boiling neutral permanganate, with cold concentrated
sulfuric acid and with bromine in carbon tetrachloride. Attempts to force reaction with many of these reagents lead to
substitution of one or more of the hydrogen atoms, a reaction
rarely encountered among aliphatic compounds.

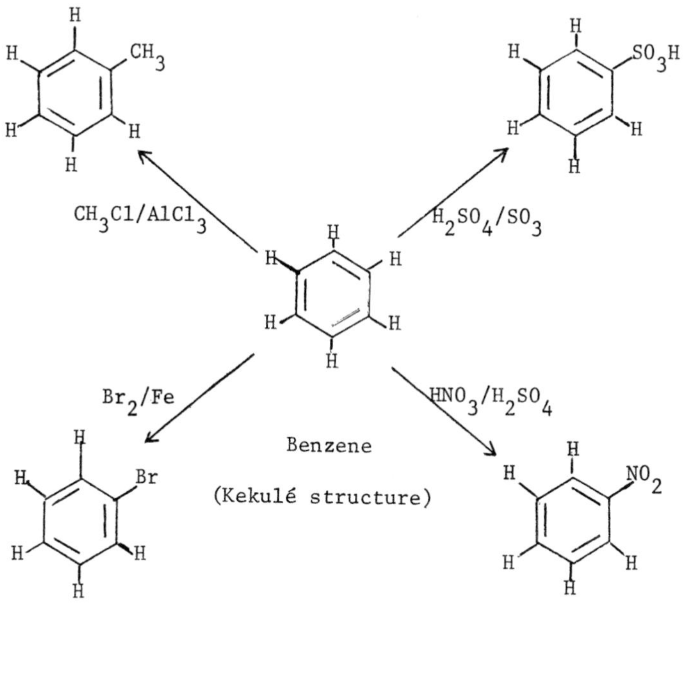

Benzene

(Kekulé structure)

2. FUNCTIONAL GROUP ORGANIC CHEMISTRY

FUNCTIONAL GROUPS

The chemical properties of organic compounds are typically those of certain characteristic groups of atoms attached to the relatively inert carbon backbones. The groups most commonly encountered are listed below (in the order in which their chemistry will be discussed) so that they may be readily committed to memory.

Amines: $-NH_2$ $>$ NH $\geqslant$ N

Carboxylic acids: $-\overset{\overset{\displaystyle O}{\|}}{C}-O-H$

Phenols: $-O-H$ attached to a benzene ring

Aldehydes and ketones: $-\overset{\overset{\displaystyle H}{|}}{C}=O$ and $>C=O$

Esters: $-\overset{\overset{\displaystyle O}{\|}}{C}-O-$

Alcohols: $-O-H$ attached to aliphatic carbon

Alkenes and alkynes: $>C=C<$ and $-C\equiv C-$

Ethers: $-O-$ attached to two carbon atoms

Alkyl and aryl halides: $-Cl$, Br, I attached to aliphatic or aromatic carbon

Additional groupings discussed here only in connection with their reactions with the groups listed above are:

Acid chlorides: $-\overset{\overset{\displaystyle O}{\|}}{C}-Cl$ Acid anhydrides: $-\overset{\overset{\displaystyle O}{\|}}{C}-O-\overset{\overset{\displaystyle O}{\|}}{C}-$

Isocyanates: $-N=C=O$ Isothiocyanates: $-N=C=S$

Sulfonyl chlorides: $-SO_2-Cl$ Grignard compounds: $-MgX$

Diazonium salts: $-N_2^+Cl^-$ Hydrazines: $-NH-NH_2$

Further groupings resulting from some of the reactions considered are:

Sulfonic acids: $-SO_3H$

Alkyl hydrogen sulfates:
     $-O-SO_3H$

Sulfonamides: $-SO_2-N\langle$

Nitro compounds: $-NO_2$

Ureas: $>N-\overset{\overset{O}{\|}}{C}-N\langle$

Thioureas: $>N-\overset{\overset{S}{\|}}{C}-N\langle$

Azo compounds: $-N=N-$

Hydrazones: $>C=N-NH-$

Anils: $>C=N-$

Oximes: $>C=N-OH$

Hydroxamic acids:
     $-\overset{\overset{O}{\|}}{C}-NH-OH$

Carbamates: $-O-\overset{\overset{O}{\|}}{C}-N\langle$

AMINES

The amines may be regarded as the organic derivatives of
ammonia, and many of their reactions simulate those of that
simple inorganic compound:

$$H-\underset{\overset{|}{H}}{\overset{\overset{|}{H}}{N}}|
\qquad CH_3-\underset{\overset{|}{H}}{\overset{\overset{|}{H}}{N}}|
\qquad CH_3-\underset{\overset{|}{CH_3}}{\overset{\overset{|}{H}}{N}}|
\qquad CH_3-\underset{\overset{|}{CH_3}}{\overset{\overset{|}{CH_3}}{N}}|$$

Ammonia   Methylamine   Dimethylamine   Trimethylamine

          (Primary   (Secondary aliphatic) (Tertiary aliphatic)
          aliphatic)

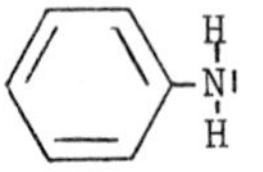

2. FUNCTIONAL GROUP ORGANIC CHEMISTRY

Aniline (phenylamine)

(Primary aromatic)

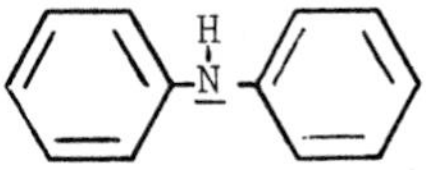

Diphenylamine

(Secondary aromatic)

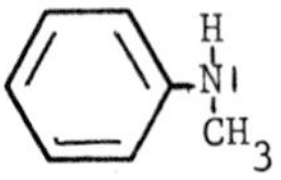

N-Methylaniline

(Secondary mixed)

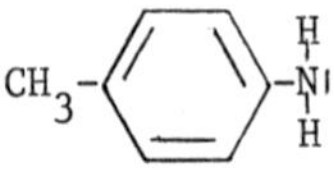

p-Toluidine

(Primary aromatic)

Basicity

All amines, like ammonia, are basic. Each, by virtue of the unshared electron-pair on the nitrogen atom, will accept a hydrogen ion from a suitable acid to form an ionised salt.

$$CH_3-CH_2-\overset{H}{\underset{H}{N}} \; + \; H_3O^+Cl^- \; \rightleftharpoons \; CH_3-CH_2-\overset{H}{\underset{H}{N}}{}^+-H \;\; Cl^- \; + \; H_2O$$

Ethylamine	Hydrochloric	Ethylamine hydrochloride
b.p. 17°*	Acid	(Ethylammonium chloride)
		m.p. 109–110°

The organic groups affect the basicity of the amine consider-ably. Aliphatic amines are more basic than ammonia and, since the common ones are readily soluble in water, they turn indica-tors to the basic color (red litmus to blue, e.g.). Most aromatic amines are less basic than ammonia and, being only

*All temperatures are in °C

14

slightly soluble in water, often do not affect indicators.
Those having only one aromatic nucleus attached to the nitrogen
are readily soluble in dilute mineral acids (dil. HCl) and are
reprecipitated by bases (dil. NaOH). Those having two or more
benzene rings attached to nitrogen, or having nitro or other
strongly electron-attracting groups on one ring, are very weak
bases and dissolve only in fairly concentrated acids (concd.
HCl; 30-40% H_2SO_4) although they are still much more basic than
most oxygen compounds (ethers, e.g.). Strongly basic amines,
especially the aliphatic ones and some of the more basic aromatic ones, may be titrated with standard solutions of mineral
acid (HCl) using indicators which change color on the acid side
(methyl red, e.g.), and equivalent weights may be thus determined.

Most salts of amines are crystalline solids and many are
useful as solid derivatives. Hydrochlorides and sulfates
usually crystallize well from concentrated hydrochloric or
30-40% sulfuric acid even though they are quite soluble in
water (common-ion effect). Organic acids often give readily
crystallized salts with the strongly basic aliphatic amines,
and picric acid (2,4,6-trinitrophenol) is particularly useful,
especially with tertiary amines which, having no replaceable
hydrogen, are not readily converted into amides.

2. FUNCTIONAL GROUP ORGANIC CHEMISTRY

Dimethylaniline   Picric acid   Dimethylaniline picrate
b.p. 193°   m.p. 162°

Solid picrates are formed even by very weakly basic amines
(diphenylamine, e.g.). These are best formulated as molecular
complexes rather than as salts.

Amides

Primary and secondary amines, like ammonia, are readily
converted into their acyl derivatives, the amides. Some amides
may be prepared by heating carboxylate salts of the amines but
this usually is a poor preparative procedure since the salt

Aniline benzoate   Benzanilide
m.p. 163°

dissociates extensively at the dehydration temperature and much
is lost by volatilization of the acid and the amine. Acylation
is better effected by acid derivatives such as anhydrides,
chlorides and esters.

Aniline   Acetic anhydride   Acetanilide
b.p. 183°   m.p. 114°

Butylamine   Benzoylchloride   N-Butylbenzamide

Methylamine   Methyl salicylate   N-Methyl salicylamide
                                       m.p. 91°

Acylation by means of an acid chloride or anhydride usually is best effected in the presence of a base and often is carried out by shaking a suspension of the amine with excess of the acid derivative in aqueous sodium hydroxide (Schotten-Baumann procedure) or by heating the reactants with pyridine. Acylation by esters needs no base and gives excellent yields of readily purified products but is slow and is best effected by allowing a mixture of reactants to stand at room temperature for several days. It is catalyzed by pyridine.

17

2. FUNCTIONAL GROUP ORGANIC CHEMISTRY

Amides usually are most excellent solid derivatives, those derived from aromatic acids (benzoic, e.g.) being most useful for aliphatic amines, and those from both aromatic and aliphatic acids (benzoic, acetic) being suitable for aromatic amines. Tertiary amines do not form amides since they have no replaceable hydrogen on the nitrogen atom.

Sulfonamides are the corresponding derivatives of amines with sulfonic acids. They cannot be prepared by heating the salts and are not conveniently prepared from the esters or anhydrides, but result from reaction of the amine with the sulfonyl halide under Schotten-Baumann conditions or in pyridine.

Aniline   Benzenesulfonyl   N-Phenyl benzenesulfonamide
                 chloride   m.p. 110°

Benzenesulfonyl chloride and p-toluenesulfonyl chloride are commonly used and give convenient results with both aliphatic and aromatic amines. Sulfonamides often are readily prepared from amines (nitroamines, e.g.) which do not readily give carboxamides.

Sulfonamides (but not carboxamides) derived from primary amines are weakly acidic, dissolve readily in sodium hydroxide solution and are reprecipitated upon acidification. Herein reside both a ready diagnosis and a separation (the Hinsberg

$$R-\underset{\underset{O}{|}}{\overset{\overset{H}{|}}{N}}-\underset{O}{\overset{O}{S}}-R' \underset{\underset{H_3O^+}{}}{\overset{OH^-}{\rightleftharpoons}} R-\underset{\underset{O}{|}}{\overset{\overset{\_}{N}}{N}}-\underset{O}{\overset{O}{S}}-R' + H_2O$$

separation) of primary, secondary and tertiary amines: The

derivative of the primary amine dissolves in the strongly alka-

line (30% NaOH) Schotten-Baumann reaction mixture and may be

separated by centrifugation or filtration from any secondary

amine derivative and unreacted tertiary amine. It is detected,

precipitated and recovered upon acidification of the strongly

alkaline supernate or filtrate. The tertiary amine may be

separated from the sulfonamide of the secondary amine by dis-

solving it in dilute acid, and is recovered upon rendering the

acid extract basic.

Half-amides result from reaction of primary and secondary

amines with the cyclic anhydrides of dicarboxylic acids. Those

derived from primary amines are converted upon heating, alone

or in glacial acetic acid, into beautifully crystalline cyclic

imides.

Methylamine   N-Methyl phthalimide
                                           m.p. 134°

19

2. FUNCTIONAL GROUP ORGANIC CHEMISTRY

Reaction with Nitrous Acid

The different types of amine react quite characteristically with nitrous acid, generated in situ from sodium nitrite and either dilute hydrochloric or concentrated sulfuric acid. Primary aliphatic amines evolve nitrogen

$$CH_3-NH_2 + HNO_2 \longrightarrow CH_3OH + N_2 + H_2O$$

The gas evolution may be difficult to distinguish from that resulting from sodium nitrite and the acid and is best observed when a solution of sodium nitrite in 80-90% sulfuric acid is added to the amine in similarly concentrated acid. Primary aromatic amines, in contrast, give diazonium salts when their

$$\text{Aniline} \quad \begin{array}{c} \boxed{}-NH_2 + HNO_2 + HCl \end{array} \longrightarrow \begin{array}{c} \boxed{}-N_2^+ \; Cl^- + 2H_2O \end{array}$$

        Aniline   Benzenediazonium chloride

cold (0-5°) solutions in dilute hydrochloric acid are treated with aqueous sodium nitrite solution. The diazonium salt may be recognized by adding the resulting solution to a solution of a phenol (phenol, β-naphthol, e.g.) in excess of dilute sodium hydroxide, when a yellow or red azo dye results.

Benzenediazonium   β-Naphthol   Benzeneazo-β-naphthol
    chloride   (orange precipitate)

Secondary amines (aliphatic and aromatic) yield N-nitrosoamines
which separate as yellow oils or low-melting solids if the
amine was aromatic or higher-aliphatic.

$$(n-C_4H_9)_2N-H + HNO_2 \xrightarrow[H_2O]{HCl} (n-C_4H_9)_2N-N=O$$

Di-n-butylamine   N-Nitroso-di-n-butyl-
                               amine (yellow oil)

Aromatic N-nitrosoamines undergo rearrangement in alcoholic
hydrochloric acid into o-or p-C-nitrosoamines which are relatively high melting.

N-Nitroso-N-methylaniline   p-Nitroso-N-methylaniline
     (yellow oil)   m.p. 118°

Tertiary aliphatic amines do not react with nitrous acid. They
dissolve in the acidic reaction mixture but are recovered on
neutralization.

21

2. FUNCTIONAL GROUP ORGANIC CHEMISTRY

Aromatic tertiary amines undergo C-nitrosation.

Dimethylaniline   p-Nitrosodimethylaniline
m.p. 87° (hydrochloride,
m.p. 177°)

Reaction with Carbonyl Compounds

Primary amines react with aldehydes and ketones, particularly when catalytic amounts of acid are present. If either reactant is aromatic, the product may be a Schiff base; particularly good results are obtained with aromatic amines and salicylaldehyde.

Aniline   Salicylaldehyde   Salicylideneaniline
m.p. 51°

Both primary and secondary amines react with isocyanates and isothiocyanates to give crystalline substituted ureas.

p-Toluidine   Phenylisocyanate   N-Phenyl-N'-p-tolylurea
m.p. 226°

22

Dimethylamine   α-Naphthylisothiocyanate   N,N-Dimethyl-N'-
α-naphthyl thiourea
m.p. 168°

Aromatic Amines; Ring Substitution

Benzenoid amines react extremely readily with electrophilic reagents. Substitution occurs ortho or para to the amino group. Polysubstitution is usual and separation of pure products often is difficult unless an excess of reagent is used; then oxidation of the amine often occurs.

p-Toluidine   2,6-Dibromo-p-toluidine
m.p. 79°

Diphenylamine   p,p'-Dinitrodiphenylamine
m.p. 216°

23

2. FUNCTIONAL GROUP ORGANIC CHEMISTRY

CARBOXYLIC ACIDS

The carboxylic acids may be regarded as derivatives of carbonic acid in which one of the OH groups is replaced by an organic group.

$$H-O-\underset{\underset{O}{\|}}{C}-O-H \qquad CH_3-\underset{\underset{O}{\|}}{C}-O-H \qquad CH_3-CH_2-\underset{\underset{O}{\|}}{C}-O-H$$

Carbonic acid   Acetic acid   Propionic acid

                        (Ethanoic acid)   (Propanoic acid)

Benzoic acid   Salicylic acid   p-Toluic acid

                 (o-Hydroxybenzoic acid) (p-Methylbenzoic acid)

Carboxylic acids are much more stable than carbonic acid and lose carbon dioxide only upon strong heating ($25\text{U}°$, upon boiling in quinoline with copper powder, upon fusing their salts with sodium hydroxide, or upon pyrolysis of their alkaline earth salts.

2-Furoic acid   Furan

$$CH_3-COO^-Na^+ \; + \; Na^+OH^- \xrightarrow{\text{fuse}} CH_4 + Na_2CO_3$$

Sodium acetate   Methane

24

$$Ba^{++}(CH_3COO^-)_2 \xrightarrow{\text{pyrolyse}} \underset{\substack{\| \\ \\}}{CH_3-\overset{O}{C}-CH_3} + BaO + CO_2$$

Barium acetate   Acetone

Some acids are decarboxylated rather more readily.

$$\underset{\substack{| \\ CH_2 \\ | \\ COOH}}{COOH} \xrightarrow{140-150°} \underset{\substack{| \\ COOH}}{CH_3} + CO_2$$

Malonic acid   Acetic acid

Acidity

All carboxylic acids, like carbonic acid, are acidic.
Each will donate a proton to a suitable base to form an ionised
salt.

$$CH_3-CH_2-CH_2-COOH + OH^- Na^+ \xrightarrow{H_2O} CH_3-CH_2-CH_2-COO^- Na^+ + H_2O$$

Butyric acid   Sodium butyrate

Most acids, if sufficiently soluble in water, will turn indicators to the acid color (blue litmus to red, e.g.) and will
liberate carbon dioxide gas from sodium bicarbonate.

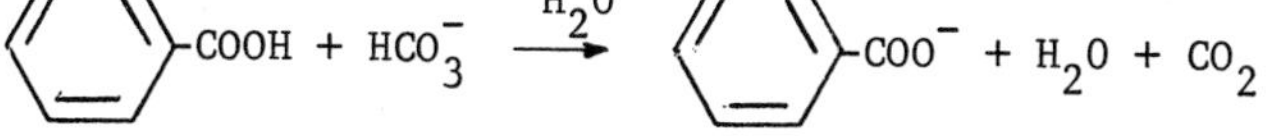

$$\bigcirc\!\!-COOH + HCO_3^- \xrightarrow{H_2O} \bigcirc\!\!-COO^- + H_2O + CO_2$$

Benzoic acid   Benzoate ion

25

2. FUNCTIONAL GROUP ORGANIC CHEMISTRY

They may be titrated with standard solutions of strong bases (NaOH, Ba(OH)$_2$, e.g.) using basic-side indicators (phenolphthalein, pH 9, e.g.) and equivalent weights are thus determined.

The carboxylate salts invariably are crystalline solids but many are very soluble in water and most of them do not have sharp characteristic melting points, so that they are of little value for identification purposes. The lead salts of the higher homologs (stearate, palmitate) and salts with complex organic bases often are useful for this purpose.

$$2CH_3-(CH_2)_{16}-COO^-Na^+ + Pb^{2+}(NO_3^-)_2 \longrightarrow (CH_3-(CH_2)_{16}-COO^-)_2Pb^{2+} + 2Na^+NO_3^-$$

Sodium stearate   Lead stearate
                                   m.p. 125°

$$CH_3-COO^-Na^+ + \text{(benzyl)}-CH_2-S-\overset{\overset{+NH_2}{\|}}{\underset{NH_2}{C}}Cl^- \longrightarrow CH_3-COO^-\overset{\overset{+NH_2}{\|}}{\underset{NH_2}{C}}-S-CH_2-\text{(phenyl)} + Na^+Cl^-$$

Sodium acetate   S-Benzyliso-   S-Benzylisothiouronium
                      thiouronium chloride acetate   m.p. 136°

Amides

Conversion into amides is equally as significant a reaction for carboxylic acids as it is for amines. Successful conversion occasionally may be achieved simply by heating the ammonium or amine salt, but more reliable results are obtained if the acid is first converted into the acid chloride by reaction with phosphorus pentachloride or thionyl chloride.

26

$$CH_3-O-\!\!\left\langle\bigcirc\right\rangle\!\!-COOH + SOCl_2 \longrightarrow CH_3-O-\!\!\left\langle\bigcirc\right\rangle\!\!-\overset{\overset{O}{\|}}{C}-Cl + SO_2 + HCl$$

p-Anisic acid   p-Methoxybenzoyl chloride
(p-Methoxybenzoic acid)

$$\xrightarrow{NH_3/H_2O} \quad CH_3-O-\!\!\left\langle\bigcirc\right\rangle\!\!-\overset{\overset{O}{\|}}{C}-NH_2$$

p-Methoxybenzamide
m.p. 163°

$$CH_3-\overset{\overset{O}{\|}}{C}-OH + PCl_5 \longrightarrow POCl_3 + HCl + CH_3-\overset{\overset{O}{\|}}{C}-Cl$$

$$CH_3-\overset{\overset{O}{\|}}{C}-Cl + \left\langle\bigcirc\right\rangle\!\!-NH_2 \longrightarrow CH_3-\overset{\overset{O}{\|}}{C}-NH-\!\!\left\langle\bigcirc\right\rangle$$

Acetanilide
m.p. 112°

$$CH_3-\overset{\overset{O}{\|}}{C}-Cl + CH_3-\!\!\left\langle\bigcirc\right\rangle\!\!-NH_2 \longrightarrow CH_2-\overset{\overset{O}{\|}}{C}-NH-\!\!\left\langle\bigcirc\right\rangle\!\!-CH_3$$

Acet-p-toluidide
m.p. 153°

A most excellent reagent for preparation of amides is the

rather expensive compound dicyclohexylcarbodiimide. The re-

agent is added to a cold solution of the acid and amine in

chloroform or ethyl acetate, the insoluble dicyclohexylurea removed by centrifugation or filtration, and the amide recovered by evaporation of the solution.

o-Toluic acid   p-Anisidine   Dicyclohexylcarbodiimide

o-Tolu-p-anisidide   Dicyclohexyl urea
(N-p-Methoxyphenyl o-methylbenzamide)

Esters

An ester results from prolonged heating of an acid with excess of an alcohol and a catalytic amount of anhydrous mineral acid (HCl, H_2SO_4) or, more readily, upon reaction of an acid chloride with an alcohol or a phenol in pyridine or aqueous base (Schotten-Baumann procedure for esters of phenols).

m-Toluic acid   Ethyl m-toluate
                                      b.p. 227°

28

Benzoyl chloride   Phenol   Phenyl benzoate
                                     m.p. 69°

Most simple esters are liquids or low-melting solids. Beautifully crystalline esters result from reaction of the sodium
salts of acids with the highly reactive α-halogeno ketones,
p-bromophenacyl bromide and p-phenylphenacyl bromide.

Sodium acetate   p-Bromophenacyl   p-Bromophenacyl acetate
                  bromide   m.p. 85°

Sodium succinate   p-Phenylphenacyl bromide

                      Di-p-phenylphenacyl succinate
                      m.p. 208°

29

2. FUNCTIONAL GROUP ORGANIC CHEMISTRY

Dicarboxylic Acids

Acids having two carboxyl groups on adjacent carbon atoms lose water and form the cyclic anhydrides upon heating.

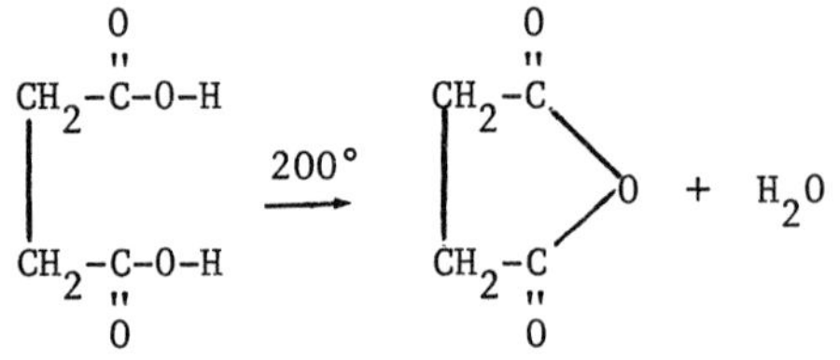

$$\begin{array}{ccc}
\underset{O}{CH_2-\overset{O}{C}-O-H} & \xrightarrow{200°} & \underset{O}{CH_2-\overset{O}{C}} \\
CH_2-C-O-H & & CH_2-C
\end{array} \quad O + H_2O$$

Succinic acid	Succinic anhydride
m.p. 185°	m.p. 120°

Phthalic acid (with benzene ring, two C-O-H groups) $\xrightarrow{200°}$ Phthalic anhydride + H_2O

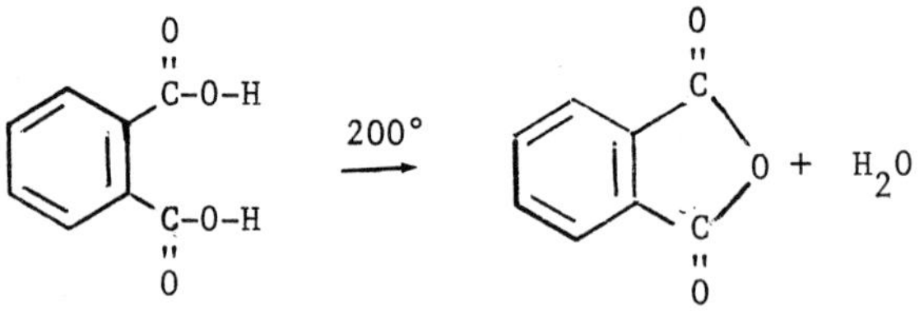

Phthalic acid	Phthalic anhydride
m.p. ca. 200°	m.p. 132°

$$O=\overset{|}{C}-O-H$$
$$CH=CH$$
$$\overset{|}{H-O-C}=O$$

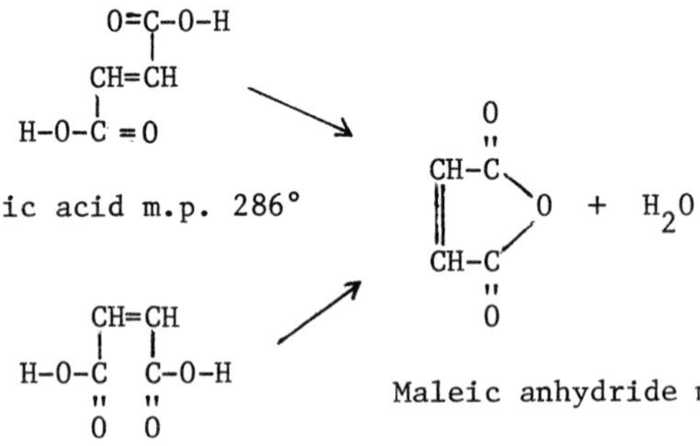

Fumaric acid m.p. 286°

$$\underset{O}{CH-\overset{O}{C}} \qquad O + H_2O$$
$$CH-C$$

Maleic anhydride m.p. 56°

$$CH=CH$$
$$H-O-\overset{|}{C} \quad \overset{|}{C}-O-H$$
$$O \quad O$$

Maleic acid m.p. 135°

The dehydration often proceeds so smoothly that the product re-quires no purification. In such cases (phthalic acid, e.g.)

characteristic identification is obtained simply by determining the m.p. of the acid and then heating the melt at ~200° for several minutes, cooling until resolidified, and redetermining the m.p. which is now that of the anhydride.

The cyclic anhydrides readily give half-amides of the diacids upon treatment with ammonia or amines, and these, upon heating, give the cyclic imides just as characteristically as the acids give the cyclic anhydrides.

Phthalic anhydride m.p. 132°   Phthalamic acid m.p. 148-9°   Phthalimide m.p. 177°

Succinic anhydride m.p. 120°   Aniline   N-Phenyl succinamic acid m.p.148.5°   N-Phenyl succinimide m.p. 156°

2. FUNCTIONAL GROUP ORGANIC CHEMISTRY

Many diacids having the carboxyl groups separated by three carbon atoms also give cyclic anhydrides upon heating but these are rarely as pure as those from 1,2-diacids. However, the crude anhydrides give well crystallized half-amides and imides.

| Glutaric acid m.p. 98° | Glutaric anhydride | N-Phenyl glutaramic acid m.p. 126-7° | N-Phenyl glutarimide m.p. 144-5° |

Aromatic Acids; Ring Substitution

Benzenoid carboxylic acids undergo ring substitution moderately readily upon reaction (nitration, halogenation, e.g.) with electrophilic reagents. Reaction usually is readily arrested at monosubstitution and, unless certain other substituents ($-OH$ or $-NH_2$ meta to $-COOH$, e.g.) are present, gives the product from substitution meta to the carboxyl group.

Benzoic acid   m-Nitrobenzoic acid
                                          m.p. 141°

PHENOLS

The phenols may be regarded as derivatives of water in which one hydrogen atom is replaced by a benzene ring.

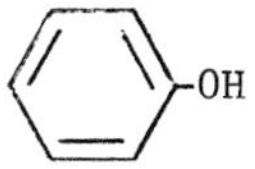

| Phenol |    p-Cresol | 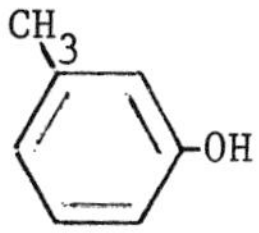   m-Cresol |
| (Hydroxybenzene) | | |

| 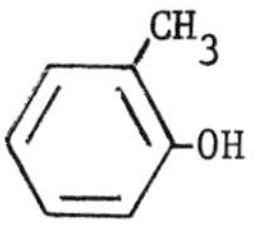 o-Cresol | 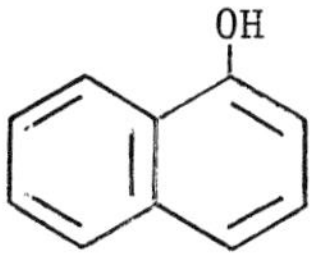 α-Naphthol | β-Naphthol |
| | (1-Hydroxynaphthalene) | (2-Hydroxynaphthalene) |

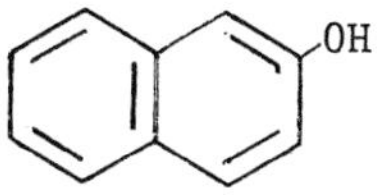

Compounds in which the oxygen atom of the hydroxyl group is not attached directly to a ring carbon atom of the aromatic ring do not show the characteristic reactions of phenols, but rather those of alcohols. Most simple phenols are slightly soluble or insoluble in water but dissolve readily in ethanol, acetone, chloroform and other organic solvents.

Acidity

The phenols are more acidic than water. They dissolve readily in dilute aqueous base (NaOH) and are reprecipitated unchanged upon acidification.

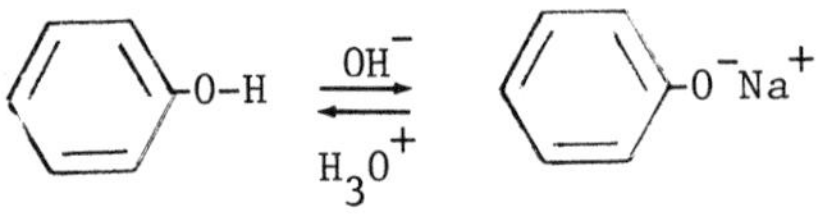

2. FUNCTIONAL GROUP ORGANIC CHEMISTRY

Simple phenols are insufficiently acidic to turn blue litmus
red or to liberate CO_2 from bicarbonate solution, however, and
so are readily distinguished from carboxylic acids. Phenols
having nitro groups or other electron-attracting substituents
on the ring are more strongly acidic, some sufficiently so to
turn indicators and to liberate CO_2 from bicarbonate.

Phenols give brilliantly colored (red, blue, green, purple)
complexes with ferric salts in water or, often more reliably,
in non-aqueous solvents (chloroform containing a trace of pyridine).

Reactions of the Hydroxyl Group

Esterification. Phenols react with acid chlorides or anhydrides in pyridine or under Schotten-Baumann conditions to
give esters, many of which are satisfactory crystalline derivatives.

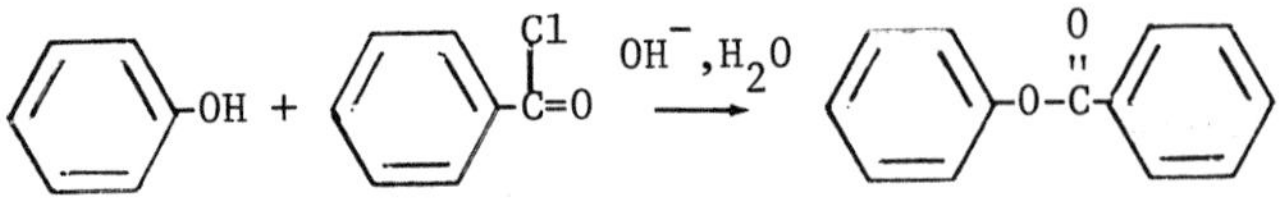

Phenol

Phenyl benzoate
m.p. 69°

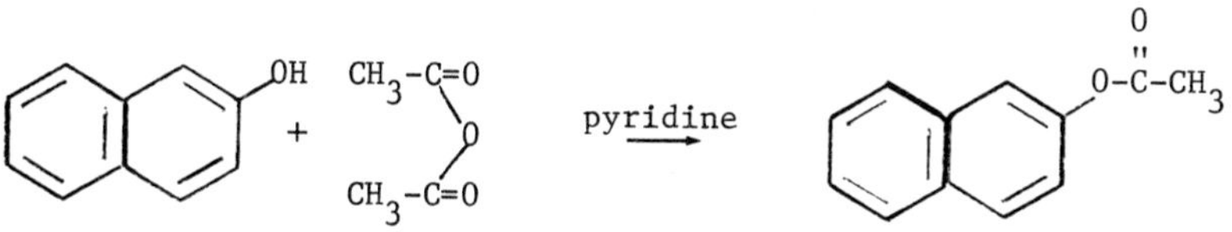

β-Naphthol

β-Naphthyl acetate
m.p. 72°

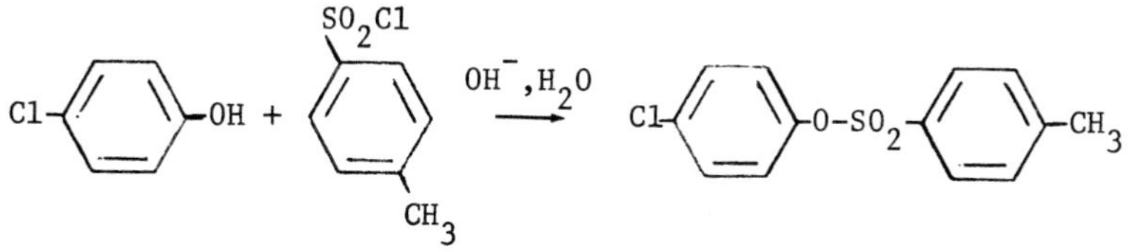

p-Chlorophenol

p-Chlorophenol p-toluene-
sulfonate m.p. 71°

With isocyanates in aprotic solvents phenols give esters
of substituted carbamic acids.

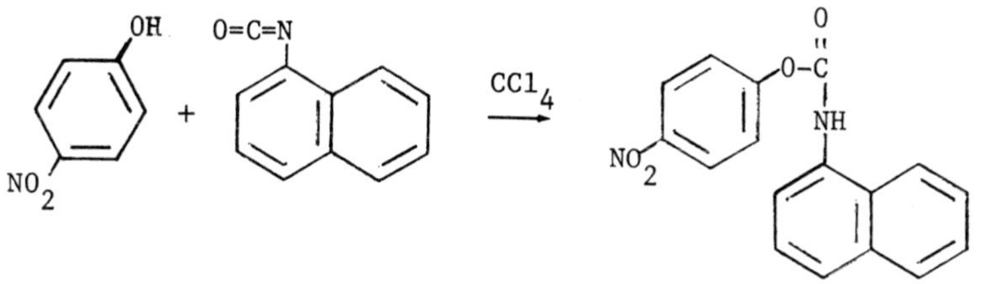

α-Naphthyl isocyanate

p-Nitrophenyl
N-α-naphthyl carbamate
m.p. 151°

Etherification. Phenoxides (solutions of phenols in

caustic alkali) give ethers, some of which are crystalline,

upon reaction with alkyl or activated aryl halides.

35

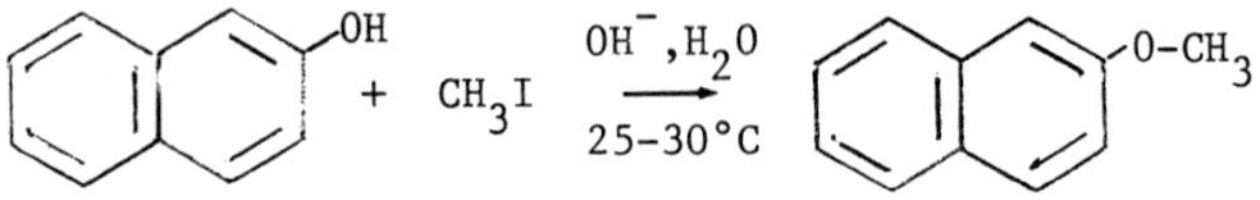

Methyl-2-naphthyl
ether (nerolin)
m.p. 72°

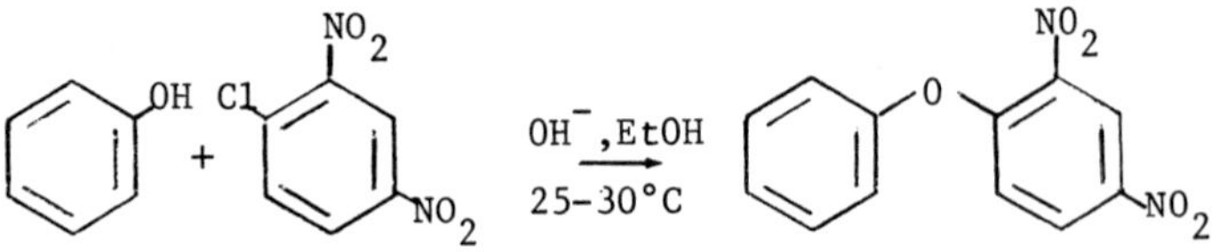

2,4-Dinitrodiphenyl ether
m.p. 69°

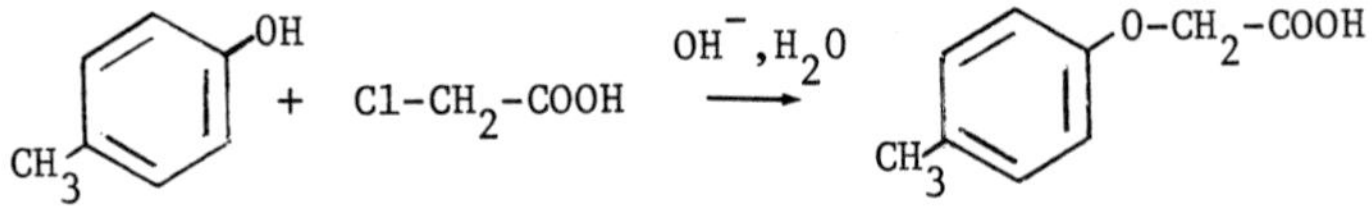

Chloracetic acid   p-Cresoxyacetic acid
                                      m.p. 136°

Ring Substitution Reactions

Phenols and their salts are extremely prone to electro-

philic substitution. Reaction occurs para and/or ortho to the

hydroxyl group and polysubstitution is common unless the incom-

ing substituent deactivates the ring and the conditions are

mild.

Azo Coupling. Phenoxides give brightly colored azo dyes

with diazonium salts,

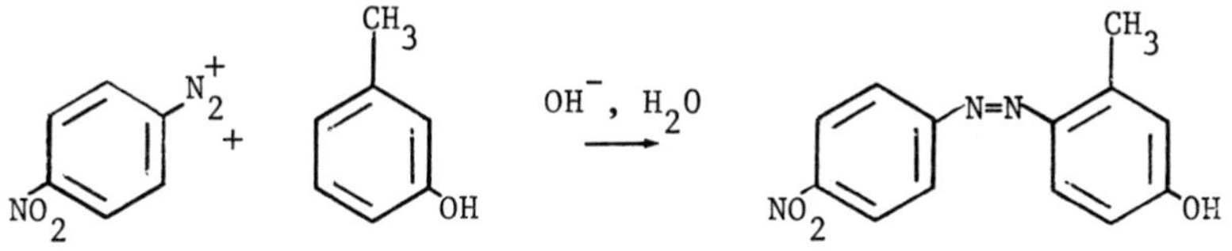

p-Nitrobenzeneazo-m-
cresol (red precipitate)

The coupling reaction is as specific and sensitive a reaction

for phenols as it is for primary aromatic amines.

Halogenation. Reaction with excess halogen at room tem-

perature usually leads to substitution at all unsubstituted

positions o or p to the hydroxyl group.

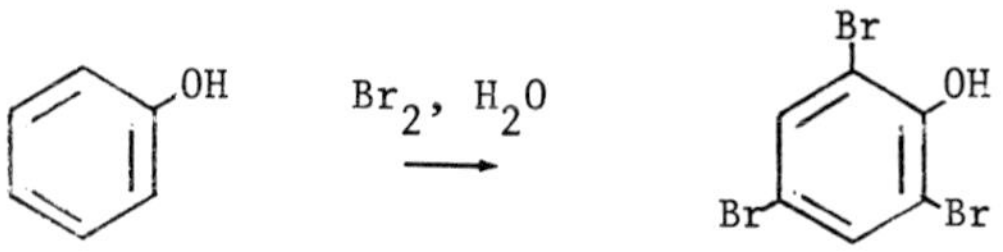

2,4,6-Tribromophenol
m.p. 95°

Use of insufficient halogen often gives difficultly separable

mixtures of mono- and polysubstituted phenols.

Nitration. Reaction with nitric acid under mild condi-

tions (30-40% HNO_3, 40-50°C) usually stops at monosubstitution

but, if both o and p positions are available, a mixture re-

sults whose separation is tedious.

2. FUNCTIONAL GROUP ORGANIC CHEMISTRY

p-Nitrophenol   o-Nitrophenol

If the p position is already occupied, as with p-cresol, p-chlorophenol or ß-naphthol, the reaction conveniently yields readily purified solid derivatives.

2-Nitro-4-chlorophenol
m.p. 87°

1-Nitro-2-naphthol
m.p. 103°

ALDEHYDES AND KETONES

These compounds contain the carbonyl group, $-\overset{|}{C}=O$, as do the carboxylic acids and their derivatives (esters, acid chlorides, amides), but with hydrogen or carbon atoms <u>only</u> attached to the carbonyl carbon atom.

Typical aldehydes:

$$\underset{\text{H}}{\overset{\text{H}}{}}\quad H-\overset{\text{H}}{\underset{|}{C}}=O \qquad CH_3-\overset{\text{H}}{\underset{|}{C}}=O \qquad C_6H_5-\overset{\text{H}}{\underset{|}{C}}=O \qquad CH_3-CH=CH-\overset{\text{H}}{\underset{|}{C}}=O$$

Formaldehyde	Acetaldehyde	Benzaldehyde	Crotonaldehyde
(Methanal)	(Ethanal)		

Typical ketones:

$$CH_3-\overset{CH_3}{\underset{|}{C}}=O \qquad CH_3-CH_2-\overset{CH_3-CH_2}{\underset{|}{C}}=O \qquad C_6H_5-\overset{CH_3}{\underset{|}{C}}=O$$

Acetone	Diethyl ketone	Acetophenone	<u>Cyclo</u>hexanone
(Propanone)	(3-pentanone)		

Aldehydes have at least one hydrogen atom attached to the carbonyl carbon atom while ketones have none.

Inorganic chemistry offers no simple counterpart to the aldehydes and ketones; thus, while amines, phenols and carboxylic acids may be regarded as the organic derivatives of ammonia, water and carbonic acid, respectively, no such analogy exists for these carbonyl compounds.

2. FUNCTIONAL GROUP ORGANIC CHEMISTRY

Reactions of the Carbonyl Group

With Primary Amino Compounds. Carbonyl compounds react readily with primary amines. The first step in the reaction involves coordination of the unshared electron pair of the amino nitrogen with the carbonyl carbon atom followed by proton transfer from nitrogen to oxygen.

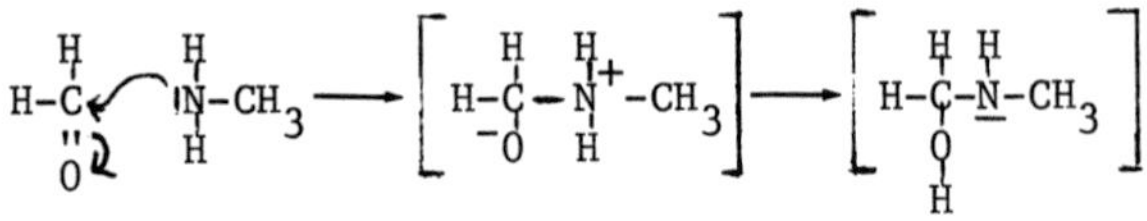

The resulting gem. hydroxyamine usually is unstable and loses water readily. The final products from simple amines and simple aldehydes or ketones are complex cyclic or linear polycondensation products having little value for characterization or identification purposes. Hydroxyamines derived from aromatic aldehydes and aromatic primary amines, however, usually eliminate water intramolecularly to give well-crystallized anils.

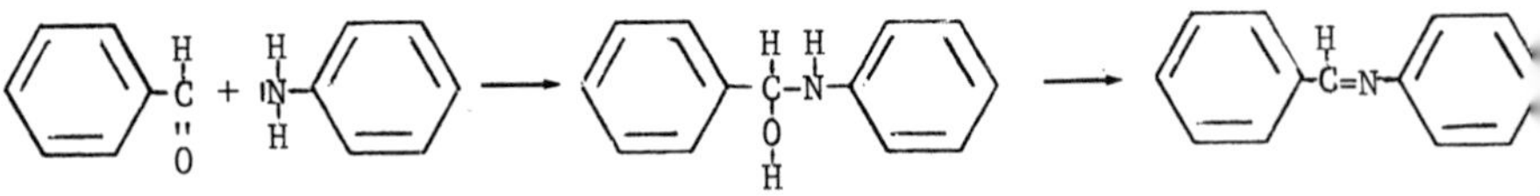

Benzylideneaniline
m.p. 54°

Even more reliable and useful for the production of solid derivatives having predictable structures is reaction with amino

40

compounds such as hydroxylamine, phenylhydrazine, semicarbazide

and 2,4-dinitrophenylhydrazine.

$$
\text{C}_6\text{H}_5\overset{\overset{\text{CH}_3}{|}}{\underset{\underset{\text{O}}{\|}}{\text{C}}} \quad + \quad \overset{\text{H}}{\underset{\text{H}}{\text{N}}}\text{-OH} \quad \longrightarrow \quad \text{C}_6\text{H}_5\overset{\overset{\text{CH}_3}{|}}{\text{C}}=\text{N-OH}
$$

Hydroxylamine   Acetophenone oxime
m.p. 59°

$$
\text{CH}_3\text{-CH}_2\text{-}\overset{\overset{\text{CH}_3}{|}}{\underset{\underset{\text{O}}{\|}}{\text{C}}} \quad + \quad \overset{\text{H}}{\underset{\text{H}}{\text{N}}}\text{-}\overset{\text{H}}{\text{N}}\text{-C}_6\text{H}_5 \quad \longrightarrow \quad \text{CH}_3\text{-CH}_2\text{-}\overset{\overset{\text{CH}_3}{|}}{\text{C}}\text{=N-}\overset{\text{H}}{\text{N}}\text{-C}_6\text{H}_5
$$

Phenylhydrazine   Methyl ethyl ketone
phenylhydrazone

$$
\text{CH}_3\text{-}\overset{\overset{\text{H}}{|}}{\underset{\underset{\text{O}}{\|}}{\text{C}}} \quad + \quad \overset{\text{H}}{\underset{\text{H}}{\text{N}}}\text{-NH-}\overset{\overset{\text{O}}{\|}}{\text{C}}\text{-NH}_2 \quad \longrightarrow \quad \text{CH}_3\text{-}\overset{\overset{\text{H}}{|}}{\text{C}}\text{=N-NH-}\overset{\overset{\text{O}}{\|}}{\text{C}}\text{-NH}_2
$$

Semicarbazide   Acetaldehyde semicarbazone
m.p. 163°

$$
\text{C}_4\text{H}_9\text{-}\overset{\overset{\text{H}}{|}}{\underset{\underset{\text{O}}{\|}}{\text{C}}} \quad + \quad \overset{\text{H}}{\underset{\text{H}}{\text{N}}}\text{-NH-C}_6\text{H}_3(\text{NO}_2)_2 \quad \longrightarrow \quad \text{C}_4\text{H}_9\text{-}\overset{\overset{\text{H}}{|}}{\text{C}}\text{=N-NH-C}_6\text{H}_3(\text{NO}_2)_2
$$

2,4-Dinitrophenyl-
hydrazine   n-Butyraldehyde 2,4-
dinitrophenylhydrazone
m.p. 123°

Hydroxylamine, phenylhydrazine and semicarbazide react smoothly

in neutral solution. In contrast, reaction with 2,4-dinitro-

phenylhydrazine proceeds even in strongly acidic medium and is

41

so characteristic that the formation of an orange crystalline precipitate with this reagent is used as a diagnostic test for aldehydes and ketones.

    Reaction with Carbanions and "Active Methylene" Compounds. Aldehydes and ketones react readily with cyanide ion; the cyanide carbanion coordinates with the carbonyl carbon atom.

$$CH_3-\underset{\underset{O}{\parallel}}{C} \quad C\equiv N \longrightarrow CH_3-\underset{\underset{O}{\mid}}{\overset{\overset{CH_3}{\mid}}{C}}-C\equiv N \xrightleftharpoons{H_2O} CH_3-\underset{\underset{OH}{\mid}}{\overset{\overset{CH_3}{\mid}}{C}}-C\equiv N$$

Acetone
cyanhydrin

Similar reaction occurs with the carbanions resulting from a variety of compounds ("active methylene compounds") in which a C-H bond is activated towards protolysis by one or more electron-attracting groups such as $-C=O$, $-C\equiv N$ or $-NO_2$.

$$HO^- \quad H-\underset{\mid}{\overset{\mid}{C}}-NO_2 \longrightarrow HO-H + {}^-\underset{\mid}{\overset{\mid}{C}}-NO_2$$

$$-\underset{\underset{O}{\parallel}}{C} \quad {}^-\underset{\mid}{\overset{\mid}{C}}-NO_2 \longrightarrow -\underset{\underset{O}{\mid}}{\overset{\mid}{C}}-\underset{\mid}{\overset{\mid}{C}}-NO_2 \xrightarrow{H_2O} \underset{\underset{H}{\overset{\mid}{O}}}{\overset{\mid}{C}}-\underset{\mid}{\overset{\mid}{C}}-NO_2$$

The severity of reaction conditions and the strength of the base required depend upon the number and effectiveness of the electron-withdrawing groups. If the activated carbon atom bears two or more hydrogens, the initial product is very prone to dehydration, either during reaction, during working-up, or

upon mild treatment with acids.

$$C_6H_5-\overset{H}{\underset{O}{C}} \quad + \quad CH_3-\overset{O}{C}-O-\overset{O}{C}-CH_3 \quad \xrightarrow{\text{AcO}^-, 130°} \quad C_6H_5-\overset{H}{C}=\overset{H}{C}-COOH$$

Benzaldehyde	Acetic anhydride	Cinnamic acid m.p. 133°

$$CH_3-\overset{H}{\underset{O}{C}} + CH_3-\overset{H}{\underset{O}{C}} \quad \xrightarrow[30°]{\text{OH}^-, H_2O} \quad CH_3-\overset{H}{\underset{OH}{C}}-\overset{H}{\underset{H}{C}}-\overset{H}{C}=O \longrightarrow CH_3-\overset{H}{C}=\overset{H}{C}-\overset{H}{C}=O$$

Acetaldehyde	Aldol	Crotonaldehyde

$$C_4H_9-\overset{H}{\underset{O}{C}} \quad + \quad CH_2\underset{\overset{C-OEt}{\underset{O}{\|}}}{\overset{\overset{O}{\|}}{C-OEt}} \quad \xrightarrow[\substack{\text{Pyridine}\\30°}]{Et_2NH,} \quad C_4H_9-C=C\underset{\overset{C-OEt}{\underset{O}{\|}}}{\overset{\overset{O}{\|}}{C-OEt}}$$

n-Butyraldehyde   Diethyl malonate

If an excess of either the active methylene compound or of the

carbonyl compound is used, and steric considerations permit,

either reactant may combine with two or more molecules of the

other.

$$3 \quad H-\overset{H}{\underset{O}{C}} + CH_3-NO_2 \quad \xrightarrow[25°]{\text{OH}^-, H_2O} \quad HO-CH_2-\overset{\overset{CH_2-OH}{|}}{\underset{\underset{CH_2-OH}{|}}{C}}-NO_2$$

2. FUNCTIONAL GROUP ORGANIC CHEMISTRY

Formaldehyde   Dimedone   Formaldehyde dimethone
                                         m.p. 189°

Aldehydes and ketones react readily with Grignard reagents and similar organometallic compounds to give alcohols.

1-Phenylethanol

Inasmuch as the organometallic reagent may be regarded as a source of carbanions or incipient carbanions (upon ionisation of the C-Mg bond), this reaction is directly analgous to the other carbanion reactions cited.   Its value as a synthetic tool cannot be overemphasized; a carbon-carbon bond is formed, producing a relatively complex and reliably predictable structure from two simpler starting materials, and the hydroxyl group in the product is readily converted into any of a variety of other functional groups.

Reaction with Other Nucleophiles.   Aldehydes react readily with bisulfite ion to give α-hydroxy sulfonic acids.

The equilibrium is shifted to the right by using a large excess of bisulfite. Some ketones, particularly simple methyl ketones and cyclic ketones, also react. The solid adduct may be iso- lated. It is decomposed by adding mineral acid or base, either of which displaces the equilibrium to the left by removing bi- sulfite ion (as SO_2 or as $SO_3^=$), thus affording a method for separating aldehydes from other substances.

Other Reactions

The atoms to which the carbonyl carbon is directly at- tached are more reactive than is the rest of the molecule. Alde- hydes are oxidized rapidly by many oxidizing agents. Argentam- mine ion (silver oxide dissolved in ammonia) and cupric ion (in alkaline solution, complexed with tartrate or citrate) are especially useful diagnostically.

$$R-CHO + 2Ag(NH_3)_2^+ + 3OH^- \longrightarrow R-COO^- + 2Ag + 2NH_3 + 2H_2O$$

$$R-CHO + 2Cu^{2+} + 5OH^- \longrightarrow R-COO^- + Cu_2O + 3H_2O$$

The red-brown precipitate of cuprous oxide and the mirror of metallic silver on the walls of the test tube are quite characteristic.

2. FUNCTIONAL GROUP ORGANIC CHEMISTRY

The methyl group of methyl ketones reacts rapidly with hypo-
halite solutions (solutions of halogens in caustic alkalis).
The resulting trihaloketones break down in the alkaline reaction mixture to give the haloform (trihalomethane) and a carboxylic acid (haloform reaction).

$$CH_3-CH_2-\overset{\overset{O}{\|}}{C}-CH_3 \quad \xrightarrow{I_2/NaOH} \quad CH_3-CH_2-\overset{\overset{O}{\|}}{C}-CI_3 \quad \xrightarrow{OH^-} \quad CH_3-CH_2-COO^- + CHI_3$$

Methyl ethyl ketone   Propionate   Iodoform

$$\text{C}_6\text{H}_5-\overset{\overset{O}{\|}}{C}-CH_3 \quad \xrightarrow{OCl^-} \quad \text{C}_6\text{H}_5-\overset{\overset{O}{\|}}{C}-CCl_3 \quad \xrightarrow{OH^-} \quad \text{C}_6\text{H}_5-COO^- + CHCl_3$$

Acetophenone   Benzoate   Chloroform

Iodoform, resulting from reaction with alkaline iodine solutions, is a pale yellow solid with a characteristic appearance
and odor. Acetaldehyde, and alcohols which give methyl ketones
or acetaldehyde upon oxidation, also exhibit the haloform reaction.

ESTERS

An ester results from elimination of the elements of water
between a molecule of a carboxylic acid and one of a phenol or
an alcohol.

$$
\begin{array}{c}
\overset{O}{\underset{}{\overset{\|}{R-C}}} \cdots O-H \\
R'-O \cdots H
\end{array}
\rightleftharpoons
\begin{array}{c}
\overset{O}{\overset{\|}{R-C}} \\
R'-O
\end{array}
+ H_2O
$$

$$CH_3-\overset{\overset{O}{\|}}{C}-O-CH_2-CH_3 \qquad CH_3-O-\overset{\overset{O}{\|}}{C}\text{---}\bigcirc$$

Ethyl acetate   Methyl benzoate
b.p. 77°   b.p. 199°

Methyl salicylate   Acetylsalicylic acid
b.p. 223°   (aspirin) m.p. 135°

The common esters are neutral, colorless, water-insoluble liquids or low-melting solids. If volatile, they have pleasant fruity odors. The typical reactions of the ester group involve splitting the C-O linkage to form the alcohol and the acid or a derivative of the acid.

Hydrolysis

Addition of water to the ester linkage occurs exceedingly slowly in water, more rapidly in dilute mineral acid (10% HCl or H_2SO_4) and most conveniently in dilute or moderately concentrated caustic alkali (NaOH, KOH) or concentrated (90%) sulfuric acid. With concentrated sulfuric acid the alcohol

47

2. FUNCTIONAL GROUP ORGANIC CHEMISTRY

component invariably is destroyed but the organic acid usually

is recovered in good yield.

Alkaline hydrolysis of an ester gives the salt of the acid from

which the free acid is liberated upon acidification.

One equivalent of base is consumed for each mole of ester

hydrolyzed. The "equivalent weight" (saponification equiva-

lent) of the ester may be determined from the amount of base

neutralized by a known weight of ester during hydrolysis. This

equivalent is equal to the molecular weight of the ester or a

simple fraction of it, and so constitutes a useful physical

constant for identification purposes.

A useful qualitative test for an ester involves heating a

drop of it with a little dilute sodium hydroxide and a few

drops of phenolphthalein in aqueous alcohol. The base is con-

sumed as the ester is hydrolyzed, so that the pink color of

alkaline phenolphthalein is discharged. Many acid anhydrides

and some aldehydes (benzaldehyde) behave similarly.

Reaction with Amines

Esters react smoothly with ammonia, most primary and some secondary amines.

Methyl benzoate   Benzamide
m.p. 129°

Benzyl acetate   Aniline

Acetanilide   Benzyl alcohol
m.p. 114°

The reaction gives a useful derivative of the acid component directly. Excess of the amine may be removed by dissolving it in aqueous hydrochloric acid.

2. FUNCTIONAL GROUP ORGANIC CHEMISTRY

Many esters react with hydrazines.

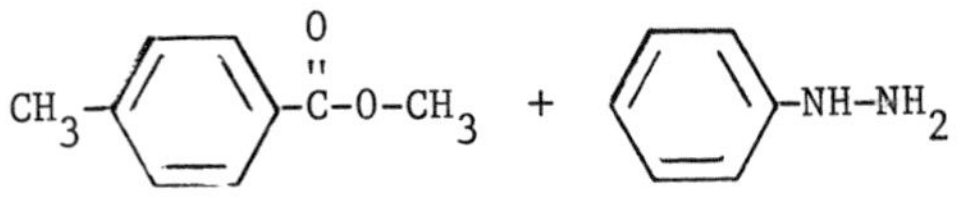

Methyl p-toluate   Phenylhydrazine

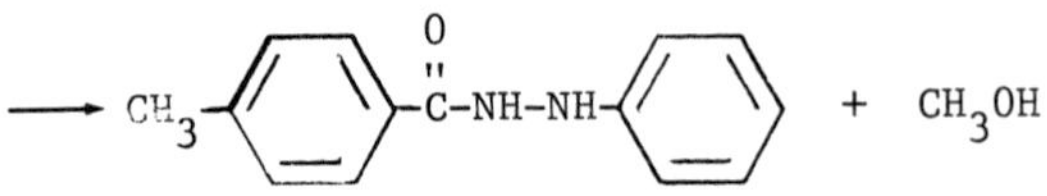

Phenyl p-toluhydrazide

The resulting hydrazide may be mistaken for the hydrazone of an aldehyde or ketone. The absence of an ester must be determined, therefore, before aldehydes and ketones are tested for.

With hydroxylamine, esters give hydroxamic acids.

$$C_3H_7-\overset{\overset{O}{\|}}{C}-O-CH_2-CH=CH_2 + NH_2OH \longrightarrow C_3H_7-\overset{\overset{O}{\|}}{C}-NH-OH + CH_2=CH-CH_2OH$$

Allyl butyrate

Hydroxamic acids give brilliant rose-red colors with aqueous ferric chloride, so that the reaction constitutes a very delicate test for esters. Phenols interfere, of course, since they too give colors with ferric chloride.

Reactions of Other Functional Groups

Esters containing other functional groups in addition to the ester function show the reactions of that group also. For example, methyl salicylate (oil of wintergreen) exhibits the characteristics of a phenol (smell; ferric chloride color).

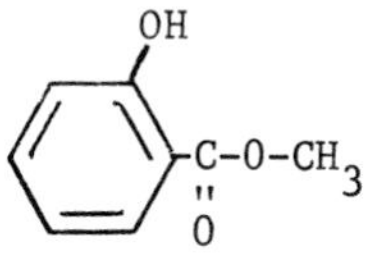

Methyl salicylate
(Oil of wintergreen)

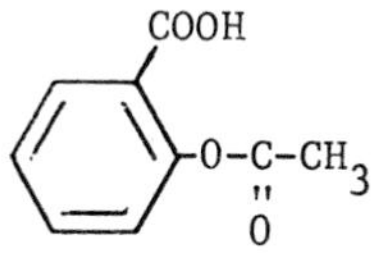

Acetylsalicylic acid
(Aspirin)

Acetylsalicylic acid (aspirin) reacts as an acid (turns indicators; liberates CO_2 from bicarbonate). Esters of phenols undergo ring-substitution reactions.

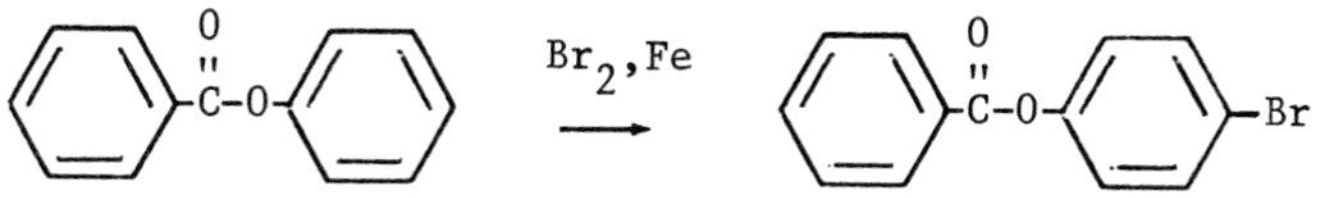

Phenyl benzoate

p-Bromophenyl benzoate

2. FUNCTIONAL GROUP ORGANIC CHEMISTRY

Alcohols

The alcohols may be regarded as derivatives of water in which one hydrogen atom is replaced by an alkyl group.

$$H-\underset{\underset{H}{|}}{\overset{\overset{H}{|}}{C}}-OH$$

Methanol
(Methyl alcohol)
(Primary)

$$H-\underset{\underset{H}{|}}{\overset{\overset{H}{|}}{C}}-\underset{\underset{H}{|}}{\overset{\overset{H}{|}}{C}}-OH$$

Ethanol
(Ethyl alcohol)
(Primary)

$$CH_3-\underset{\underset{CH_3}{|}}{\overset{\overset{CH_3}{|}}{C}}-CH_2-OH$$

Neopentyl alcohol
(Primary)

$$\underset{\underset{CH_3}{|}}{\overset{\overset{CH_3}{|}}{CH}}-OH$$

2-Propanol
(Isopropyl alcohol)
(Secondary)

$$CH_3-\underset{\underset{CH_3}{|}}{\overset{\overset{CH_3}{|}}{C}}-OH$$

1,1-Dimethylethanol
(tert -Butyl alcohol)
(Tertiary)

$$\text{(benzene ring)}-CH_2-OH$$

Phenylmethanol
(Benzyl alcohol)
(Primary)

Primary alcohols have two or more hydrogen atoms attached to the carbon atom bearing the hydroxyl group. Secondary alcohols have but one such hydrogen while tertiary alcohols have none. Hydroxy compounds containing benzene rings behave as alcohols rather than as phenols as long as the hydroxyl group is not attached directly to a ring carbon atom of the aromatic ring. Monohydric alcohols (one -OH group) containing up to three carbon atoms are miscible with water, but the water-solubility of higher alcohols decreases rapidly with increasing molecular weight.

Acidity

The alcohols are slightly less acidic than water and much less acidic than the phenols. They do not dissolve in dilute aqueous base (except, of course, those alcohols which are soluble in water alone) although they will react to some extent with <u>very</u> concentrated or solid caustic alkali.

$$CH_3-O-H + {}^-OH \, Na^+ \longrightarrow CH_3-O^- \, Na^+ + H-O-H$$

solid   solid

Salts of the alcohols (alkoxides) result upon reaction with reactive metals.

$$CH_3-CH_2-O-H + Na \longrightarrow CH_3-CH_2-O^- \, Na^+ + 1/2 \, H_2$$

Sodium ethoxide

Complete dissolution of the metal and concomitant evolution of hydrogen permit diagnosis of a lower alcohol. Higher alcohols (C_5 and above) give alkoxides which often are but slightly soluble in the alcohol and so coat the metal surface that reaction soon stops and is barely recognizable.

Oxidation

Primary and secondary alcohols are quite readily oxidized. Primary alcohols give aldehydes initially, but the latter are so readily oxidized to carboxylic acids that the yield of

53

2. FUNCTIONAL GROUP ORGANIC CHEMISTRY

aldehyde is poor unless it is removed from the reaction mixture as soon as it is formed or a considerable excess of alcohol over oxidant is used.

Benzyl alcohol   Benzaldehyde   Benzoic acid

Excellent yields of aldehydes result from dehydrogenation of simple alcohols over copper metal at 250°C.

n-Butyl alcohol   n-Butyraldehyde

Ketones result from oxidation or dehydrogenation of secondary alcohols. They resist further oxidation except under vigorous conditions.

1-Phenylethanol   Phenyl methyl ketone
                             (Acetophenone)

Tertiary alcohols are oxidized only under drastic conditions and give mixture of products.

54

Esterification

Alcohols react rapidly with cold concentrated sulfuric acid to give alkyl hydrogen sulfates.

$$CH_3-(CH_2)_{\overline{6}}CH_2OH + H_2SO_4 \xrightarrow{0°C} CH_3-(CH_2)_{\overline{6}}CH_2-O-SO_3H + H_2O$$

1-Hexanol

These half-esters are soluble both in concentrated sulfuric acid and in water, so that alcohols, even water-insoluble ones, dissolve in sulfuric acid and are not thrown out of solution upon diluting the mixture with water. They are thus readily differentiated from ethers, and from saturated and aromatic hydrocarbons. If the reaction mixture is heated (or allowed to become hot from heat of reaction or dilution) dehydration of the alcohol to alkenes, ethers and brown tarry products usually occurs.

$$CH_3-CH_2-OH \xrightarrow{hot\ H_2SO_4} H_2C=CH_2 \quad and \quad CH_3-CH_2-O-CH_2-CH_3$$

Ethanol   Ethene   Diethyl ether

Alcohols react very slowly with most carboxylic acids. The reaction is catalyzed by traces of mineral acids but still is quite slow.

2. FUNCTIONAL GROUP ORGANIC CHEMISTRY

$$CH_3-(CH_2)_2-CH_2OH + CH_3-\overset{\overset{\textstyle O}{\|}}{C}-OH \xrightarrow[\substack{reflux \\ 5\ hr}]{HCl} CH_3-(CH_2)_2-CH_2-O-\overset{\overset{\textstyle O}{\|}}{C}-CH_3 + H_2O$$

n-Butyl alcohol   Acetic acid   n-Butyl acetate

An equilibrium mixture of alcohol, acid, ester and water is

obtained. Esterification using acid chlorides or anhydrides

is rapid and complete. Solid esters result from reaction with

the chlorides or anhydrides of complex acids.

Ethanol   3,5-Dinitrobenzoyl   Ethyl 3,5-dinitrobenzoate
                 chloride   m.p. 94°

2-Propanol   3-Nitrophthalic   2-Propyl 3-nitrophthalate
               anhydride   m.p. 153°

Reaction with acetyl chloride (and other acid chlorides)

characteristically gives hydrogen chloride gas.

$$(CH_3)_3C-OH \ + \ CH_3-\overset{\overset{O}{\|}}{C}-Cl \ \longrightarrow \ (CH_3)_3C-O-\overset{\overset{O}{\|}}{C}-CH_3 \ + \ HCl$$

tert. Butyl   Acetylchloride   tert. Butyl acetate
alcohol

The characteristic pleasant fruity odor of a volatile ester derived from an alcohol of low or moderate molecular weight is
observed after destruction of excess of the reagent.

Alcohols are converted into alkyl halides by reaction with
concentrated halogen hydracids, halide salts and concentrated
sulfuric acid, phosphorus halides and thionyl chloride.

$$CH_3-(CH_2)_2-CH_2-OH \ \xrightarrow{\ NaBr, H_2SO_4\ } \ CH_3-(CH_2)_2-CH_2-Br$$

n-Butyl alcohol   n-Butyl bromide

$$(CH_3)_3C-OH \ \xrightarrow{\ conc.\ HCl,\ ZnCl_2\ } \ (CH_3)_3C-Cl$$

tert -Butyl alcohol   tert -Butyl chloride

Benzyl alcohol   Benzyl bromide

2. FUNCTIONAL GROUP ORGANIC CHEMISTRY

The reaction with halogen hydracids is similar to that with sulfuric acid; indeed, the mechanisms are very similar and quite different from that of esterification with carboxylic acids. Tertiary alcohols are much more reactive than secondary, and these more than primary alcohols. Thus a simple tertiary alcohol will react almost instantly with a solution of zinc chloride in concentrated hydrochloric acid giving the tertiary alkyl halide which separates as an upper liquid phase immiscible with the reaction mixture. A simple secondary alcohol will dissolve and produce the insoluble alkyl halide slowly upon standing, or upon warming. A primary alcohol will dissolve but reacts very slowly or not at all. The three classes of alcohol may thus be differentiated (Lucas' test).

UNSATURATED ALIPHATIC HYDROCARBONS (Alkenes or olefins and alkynes or acetylenes)

These hydrocarbons exhibit valence unsaturation; that is, they have on adjacent carbon atoms two or four electrons surplus to those required for single bonding of the carbon atoms to each other and to hydrogen. These surplus electrons are paired to form four-electron or six-electron bonds between certain of the carbon atoms. Such bonds commonly are called double and triple bonds and are represented by double or triple lines.

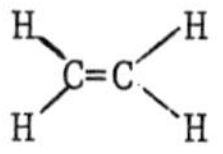

Ethene
(Ethylene)

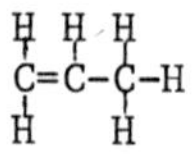

Propene
(Propylene)

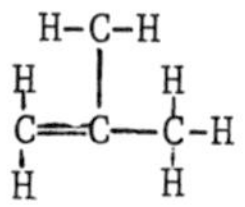

Methyl propene
(Isobutene)

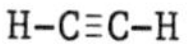

H-C≡C-H

H-C≡C-Ç-H

Butadiene

Ethyne
(Acetylene)

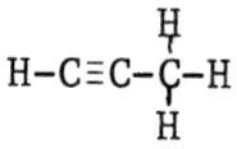

Propyne
(Methylacetylene)

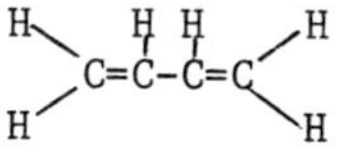

Butadiene

Multiple bonds are very reactive towards many common reagents.

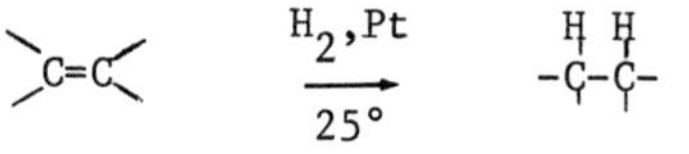

$$\text{C=C} \xrightarrow[25°]{\text{H}_2,\text{Pt}} \text{-C-C-}$$

Alkane

$$\text{C=C} \xrightarrow[0°C]{\text{H}_2\text{SO}_4} \text{-C-C-O-SO}_3\text{H}$$

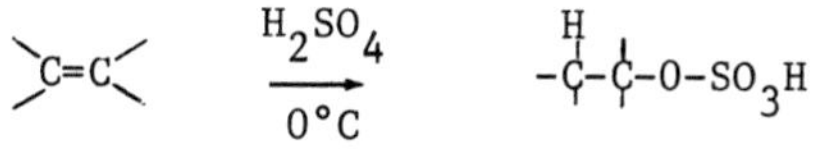

Alkyl hydrogen sulfate

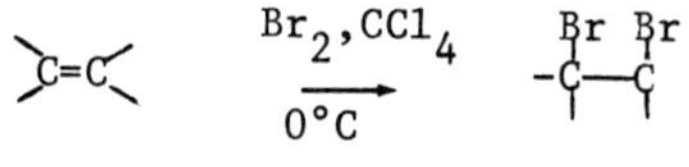

$$\text{C=C} \xrightarrow[0°C]{\text{Br}_2,\text{CCl}_4} \text{-C-C-}$$

vic. Alkylene dibromide
(Dibromoalkane)

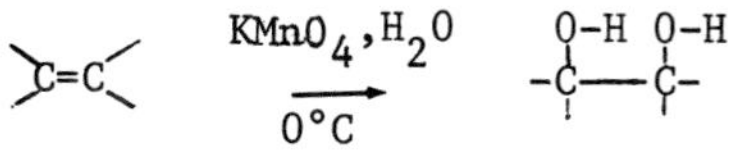

$$\text{C=C} \xrightarrow[0°C]{\text{KMnO}_4,\text{H}_2\text{O}} \text{-C-C-}$$

1,2-Diol (vic. glycol)

2. FUNCTIONAL GROUP ORGANIC CHEMISTRY

$$-C\equiv C- \quad \xrightarrow{Br_2,CCl_4} \quad -\overset{Br}{\underset{}{C}}\!\!-\!\!\overset{Br}{\underset{}{C}}- \quad \xrightarrow{Br_2,CCl_4} \quad -\overset{Br}{\underset{Br}{C}}\!\!-\!\!\overset{Br}{\underset{Br}{C}}-$$

1,1,2,2-Tetrabromoalkane

$$-C\equiv C- \quad \xrightarrow[HgSO_4,25°]{H_2O,H_2SO_4} \quad -\overset{H}{\underset{H}{C}}\!\!-\!\!\overset{}{\underset{O}{C}}\!\!-$$

Aldehyde or ketone

Such reaction, involving simple addition of the reagent to the multiple bond, almost always leaves a single bond in place of the multiple linkage. The visible characteristics of the reaction often afford easy differentiation of unsaturated hydrocarbons from their saturated or aromatic counterparts which show no such characteristics. Thus, alkenes, all of which are highly insoluble in water, dissolve in cold concentrated sulfuric acid and are not thrown out of solution upon dilution with water. They decolorize bromine in carbon tetrachloride virtually instantaneously without liberating hydrogen bromide, and, despite their low solubility in water, rapidly discharge the purple color of dilute neutral aqueous potassium permanganate solution.

The position of the multiple bond allows possibilities for isomerism in addition to those possible for the corresponding alkanes. Thus there are three alkenes C_4H_8 as opposed to two alkanes C_4H_{10}. The cycloalkenes bear to the cycloalkanes the same relationship of structure and reactivity that the alkenes bear to the alkanes.

ETHERS

The ethers may be regarded as derivatives of water in which both hydrogen atoms are replaced by carbon atoms. Either or both of the attached organic groups may be aliphatic or benzenoid.

CH_3-O-CH_3 CH_3-O-⟨benzene ring⟩   ⟨tetrahydrofuran ring structure with CH_2 groups and O⟩

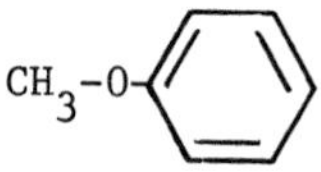

Dimethyl ether   Methyl phenyl ether   Tetrahydrofuran
                    (Anisole; methoxybenzene)

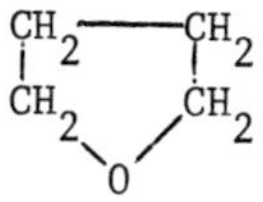

$CH_3-O-CH_2-CH_2-O-CH_3$

Ethylene glycol dimethyl ether
(1,2-Dimethoxyethane)

Reactions of the Oxygen Atom

The ethers do not exhibit the reactions of the hydrogen atoms of water, of course, but they do undergo the reactions of its oxygen atom. They accept protons from strong acids to give oxonium ions, analogous to the hydronium ion.

61

$$H-\overline{O}-H + H-A \rightleftharpoons H-\overset{\overset{H}{|}+}{\underline{O}}-H + A^-$$

$$R-\overline{O}-R + H-A \rightleftharpoons R-\overset{\overset{H}{|}+}{\underline{O}}-R + A^-$$

The protonation is reversed by bases (such as water).

$$R-\overset{\overset{H}{|}+}{\underline{O}}-R + H-\overline{O}-H \longrightarrow R-\overline{O}-R + H-\overset{\overset{H}{|}+}{\underline{O}}-H$$

Thus, **ethers,** which usually are insoluble in water, dissolve readily in cold concentrated sulfuric acid and are reprecipitated, unchanged, upon dilution of the solution with water. This behavior contrasts sharply with that of alcohols and alkenes, which are not reprecipitated upon dilution, and that of alkanes, arenes and halogenohydrocarbons, which do not dissolve.

Ethers (and all other oxygen compounds) solvate anhydrous ferric thiocyanate. They thus dissolve the greenish-black solid giving red solutions. Hydrocarbons and halogenohydrocarbons do not solvate the complex.

Fission of Ethers

Bonds from alkyl carbon atoms to the ether oxygen are broken under rather forcing conditions upon treatment with Lewis acids or concentrated protic acids. With concentrated hydriodic acid, alkyl iodides and phenols usually result from alkyl and phenyl groups, respectively, attached to the ether oxygen.

$$CH_3-CH_2-O-CH_2-CH_3 \quad \xrightarrow{HI} \quad 2CH_3-CH_2-I + H_2O$$

        Diethyl ether   Ethyl iodide

                                         Methyl   Phenol
                                         iodide

Ring-substitution of Aryl Ethers

A benzene ring attached to an ether oxygen atom is highly reactive to electrophilic substitution, although its reactivity is still far less than that of such a ring attached to a phenolic hydroxyl group. Positions para and ortho to the oxygen function are especially reactive.

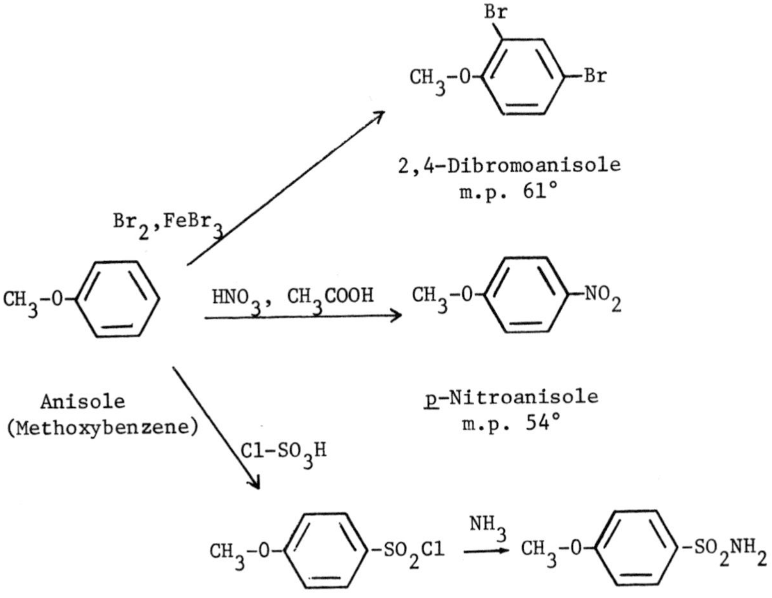

CH_3-O-⟨Br, Br⟩

2,4-Dibromoanisole
m.p. 61°

Br_2,$FeBr_3$

CH_3-O-⟨⟩ Anisole
(Methoxybenzene)

$\xrightarrow{HNO_3, \ CH_3COOH}$ CH_3-O-⟨⟩$-NO_2$

p-Nitroanisole
m.p. 54°

$Cl-SO_3H$

CH_3-O-⟨⟩$-SO_2Cl$ $\xrightarrow{NH_3}$ CH_3-O-⟨⟩$-SO_2NH_2$

p-Methoxybenzene-
sulfonyl chloride

p-Methoxybenzene-
sulfonamide m.p. 111°

Benzenoid ethers having alkyl side chains may be oxidized
to the corresponding carboxylic acids.

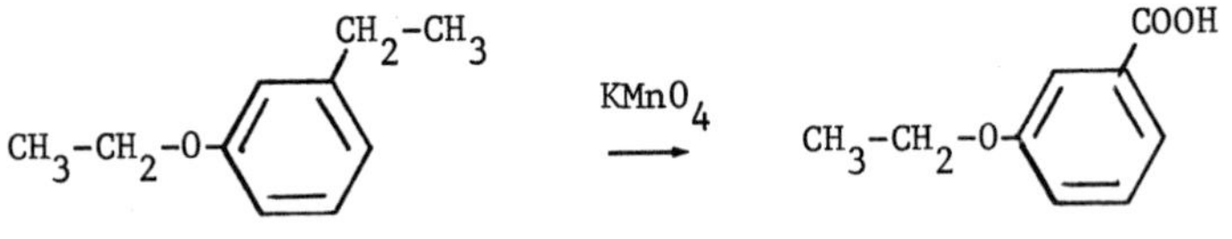

CH_3-CH_2-O-⟨CH_2-CH_3⟩ $\xrightarrow{KMnO_4}$ CH_3-CH_2-O-⟨COOH⟩

m-Ethylphenetole
(Ethyl m-ethylphenyl ether)

m-Ethoxybenzoic acid
m.p. 137°

BENZENOID HYDROCARBONS (Arenes)

Despite the valence unsaturation apparent in the structure
of benzene, which has six electrons surplus to those required
for single bonding of its carbon atoms to each other and to
hydrogen,this hydrocarbon and its derivatives are quite resis-
tant to the addition reactions characteristic of other unsaturated hydrocarbons. Reaction either does not occur or results
in replacement (substitution)of a hydrogen atom by the attack-
ing group. The first step in reaction, coordination of an
electron-deficient species (electrophile) derived from the
reagent with a pair of unsaturation electrons, apparently is
common to both arenes and alkenes,

but the tendency towards reformation of the stable benzene system is so great that the resulting electron-deficient carbonium

65

ion achieves stability by losing a proton (to a suitable

acceptor) from the carbon atom bearing the incoming group

rather than by coordination with an electron donor (nucleo-

phile, bromide ion in the above example).

Thus, simple arenes do not dissolve in cold concentrated

sulfuric acid, although the more reactive ones (mesitylene,

1,3,5-trimethylbenzene, e.g.) slowly undergo substitution

giving sulfonic acids. They are hydrogenated (H_2, Pt or Ni)

only under forcing conditions and resist reaction with perman-

ganate, dichromate and other oxidizing agents. Indeed, satu-

rated aliphatic side chains are smoothly degraded by these

oxidizing agents giving good yields of carboxylic acids.

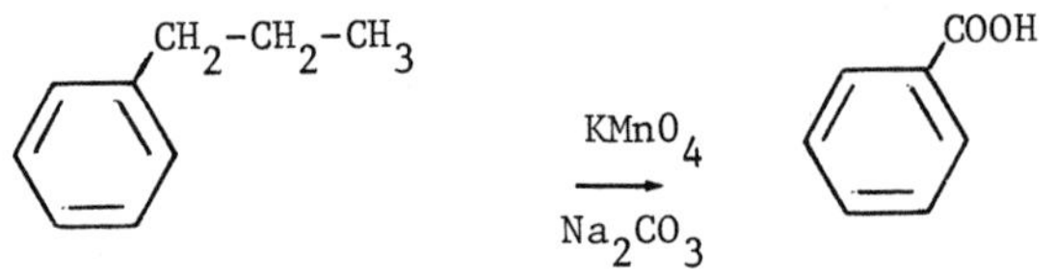

n-Butylbenzene

Benzoic acid
m.p. 121°

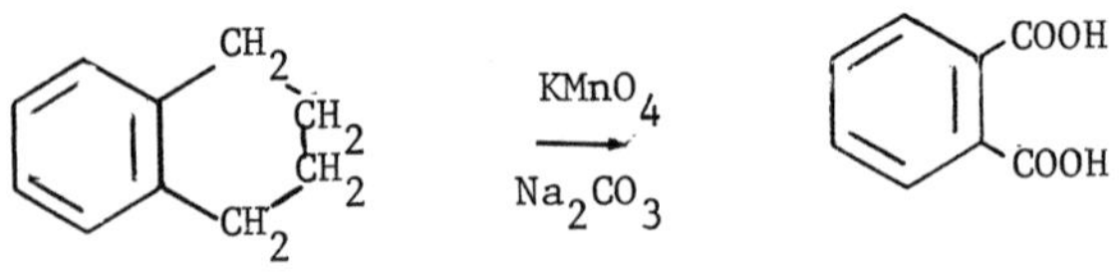

Tetralin
(Tetrahydronaphthalene)

Phthalic acid
m.p. 195-205°

Substitution Reactions of Benzene

The common aromatic substitution reactions are nitration, sulfonation and chlorosulfonation, halogenation, and reaction with a variety of acyl and alkyl compounds such as acid anhydrides, acid chlorides, alkyl halides, alkenes, alcohols etc. (Friedel-Crafts reaction).

Bromobenzene   p-Dibromobenzene

Toluene   p-Xylene

Nitrobenzene   m-Dinitrobenzene

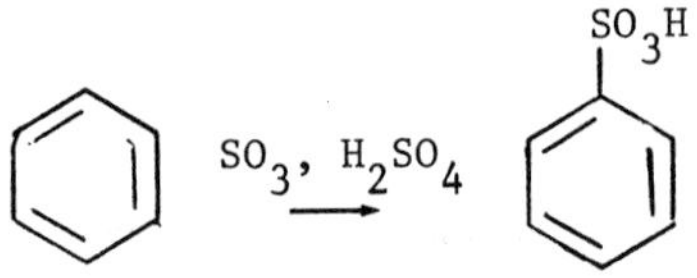

2. FUNCTIONAL GROUP ORGANIC CHEMISTRY

Benzenesulfonic acid

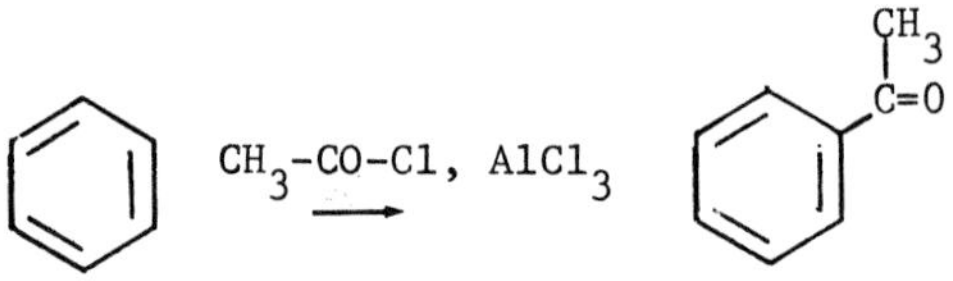

Acetophenone

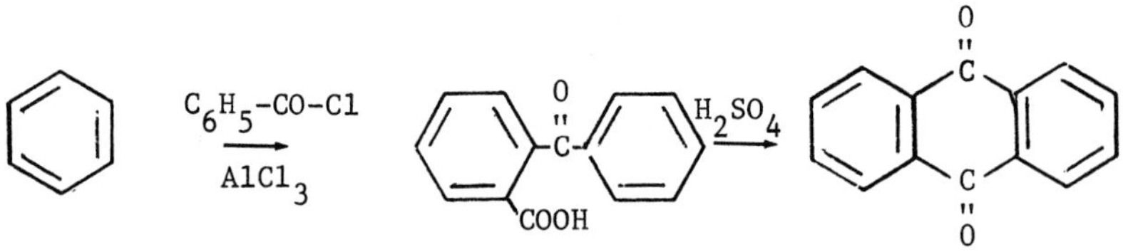

o-Benzoylbenzoic   Anthraquinone
acid

The reactions all require promotion or catalysis. Nitration

and sulfonation are promoted by concentrated sulfuric or other

anhydrous strong acids. Halogenation and the various Friedel-

Crafts reactions are promoted by Lewis acids such as anhydrous

aluminum chloride, zinc chloride, ferric bromide and boron

trifluoride. The alkyl group introduced by alkylation acti-

vates the ring to further reaction so that complex mixtures of

polysubstitution products usually result. All of the other

substituents introduced by the reactions indicated deactivate

the ring, so that reaction can be controlled to yield

predominantly the monosubstitution product. A second, and in some cases even a third nitro, halogen or sulfonate group can be introduced under more forcing reaction conditions. Acylation, however, cannot be forced beyond the monosubstitution stage except in a few very special cases (anthraquinone from o-benzoylbenzoic acid, e.g.).

Chlorosulfonation is analogous to both sulfonation and acylation. The reagent, chlorosulfonic acid, is its own catalyst.

$$\text{Excess Cl-SO}_3\text{H} \longrightarrow$$

Benzenesulfonyl chloride

The resulting sulfonyl chloride is much more readily separated from excess reagent than is a sulfonic acid from a sulfonation mixture, and reacts smoothly with ammonia and amines to give sulfonamides having predictable structures and characteristic melting points.

Substitution Reactions of Substituted Benzenes

Further substitution of a benzene derivative occurs with an ease and at a position on the ring that is governed by the substituent already present and essentially independent of the nature of the incoming group. Substituents are classified in four groups:

2. FUNCTIONAL GROUP ORGANIC CHEMISTRY

(1) <u>Deactivating</u>; meta-directing

$-N(CH_3)_3^+$ $-NO_2$ $-SO_3H$ $-C\equiv N$ $\overset{\overset{\displaystyle O}{\|}}{-C-OH}$

$\overset{\overset{\displaystyle O}{\|}}{-C-}$ $\overset{\overset{\displaystyle O}{\|}}{R-C-O-R}$ $\overset{\overset{\displaystyle O}{\|}}{-C-NH_2}$ $-NH_3^+$

example:

$Br_2, FeBr_3$

Nitrobenzene   <u>m</u>-Bromonitrobenzene

(2) <u>Weakly</u> <u>deactivating</u>; ortho-para-directing

The halogens; -Cl   -Br   -I

example:

HNO_3, H_2SO_4

Bromobenzene   <u>p</u>-Bromonitro-   <u>o</u>-Bromonitro-
                          benzene   benzene

(3) <u>Activating</u>; ortho-para-directing

Alkyl $\overset{\overset{\displaystyle O}{\|}}{-O-C-R}$ $-O-R$ $\overset{\overset{\displaystyle O}{\|}}{-NH-C-R}$

70

example:

| Methoxybenzene | p-Methoxybenzene- |
| (Anisole) | sulfonyl chloride |

(4) **Very powerfully activating**; ortho-para-directing

$-NH_2$ $-NHR$ $-NR_2$ $-OH$

example:

Phenol   p-Nitrophenol o-Nitrophenol

Reaction of a ring bearing a meta-directing substituent gives predominantly one product of predictable structure. A mixture of two products invariably results from a ring already bearing an ortho-para-directing substituent but, of the two, the para isomer almost always predominates, is higher-melting or -boiling and less soluble in solvents, and usually is fairly readily isolated.

Amines and phenols are so reactive that reaction often occurs at all available ortho and para positions.

71

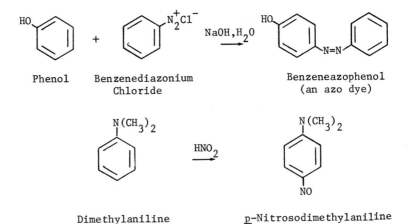

Phenol
$\xrightarrow{Br_2, H_2O}$
2,4,6-Tribromophenol
m.p. 95°

m-Toluidine
$\xrightarrow{Br_2, H_2O}$
Tribromo-m-toluidine
(3-Amino-2,4,6-tribromotoluene)
m.p. 101°

Indeed, these substances undergo a number of unique substitution reactions.

Phenol + Benzenediazonium Chloride
$\xrightarrow{NaOH, H_2O}$
Benzeneazophenol
(an azo dye)

Dimethylaniline
$\xrightarrow{HNO_2}$
p-Nitrosodimethylaniline

ALKYL HALIDES

The alkyl halides may be regarded as esters derived from the halogen hydracids and the alcohols.

$$R-OH + H-X \longrightarrow R-X + H_2O$$

Alkyl halides

Primary   Secondary   Tertiary   Allylic

CH_3-I $CH_3-\overset{Br}{\underset{}{C}H}-CH_3$ $CH_3-\overset{Cl}{\underset{CH_3}{C}}-CH_3$ $CH_2=CH-CH_2Cl$

Methyl iodide Isopropyl bromide tert-Butyl   Allyl chloride
                                  chloride

CH_3-CH_2-Br $CH_3-(CH_2)-\overset{I}{\underset{}{C}H}-CH_3$ Benzylic   Vinylic

Ethyl bromide 2-Iodooctane

$CH_3-CH_2-CH_2-CH_2-Cl$ $-CH_2-Cl$ $CH_2=CH-Br$

n-Butyl chloride   Benzyl   Vinyl
                                                      chloride   bromide

They are non-polar compounds, insoluble in water and in cold concentrated sulfuric acid. They may be distinguished from the alkanes by detection of the halogen as halide ion after fusion of the substance with sodium metal or hydrolysis with caustic alkali (sodium or potassium hydroxide solution).

Substitution Reaction of Alkyl Halides

Alkyl halides undergo a veritable plethora of reactions with reagents (nucleophiles) capable of providing a pair of

73

electrons to form a bond with the carbon atom, which loses its halide ion.

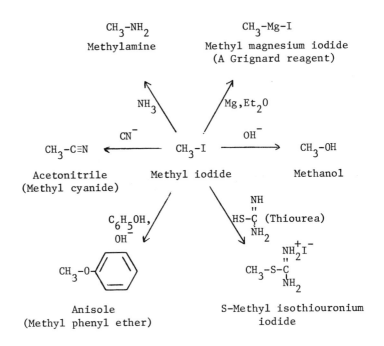

The reactivity of the alkyl halides varies widely with structure. Iodides are more reactive than bromides and these are more reactive than chlorides of the same structure. Allyl halides are most reactive, followed, in order of decreasing reactivity, by benzyl, tertiary, methyl, secondary-primary and vinyl halides. The vinyl halides, like the aryl halides, are virtually unreactive to nucleophiles. If the reagent is strongly basic (OH^-, CN^-, OR^-, e.g.) and the halide is tertiary or secondary, elimination of HX frequently predominates over substitution.

$$CH_3-\underset{\underset{CH_3}{|}}{\overset{\overset{CH_3}{|}}{C}}-Br \quad \xrightarrow{\text{OH}^-} \quad CH_3-\underset{\underset{CH_3}{}}{\overset{\overset{CH_2}{\|}}{C}}$$

tert — Butyl bromide   Isobutene

ARYL HALIDES

The aryl halides are not conveniently regarded as esters (of phenols) but rather as halogen substitution products of benzenoid hydrocarbons. They are not prepared from the phenols, nor are they hydrolyzed to them except in special cases or under special conditions. They are non-polar substances, insoluble in water and cold concentrated sulfuric acid, and are distinguished from the aromatic hydrocarbons by detection of the halogen as halide ion after fusion with metallic sodium.

They are unreactive to nucleophiles (compare the vinyl halides) but will react with magnesium to give Grignard reagents which undergo a variety of useful reactions.

Bromobenzene $\xrightarrow{\text{Mg,Ether}}$ Phenylmagnesium bromide $\xrightarrow[\text{(2) HCl,H}_2\text{O}]{\text{(1) CO}_2}$ Benzoic acid m.p. 121°

The characteristic reactions of aryl halides are the electrophilic substitution reactions (nitration, halogenation,

chlorosulfonation, etc.) of the benzene ring. The ring is
weakly deactivated by the halogen atom, and the incoming substituent is directed into the ortho and para positions.

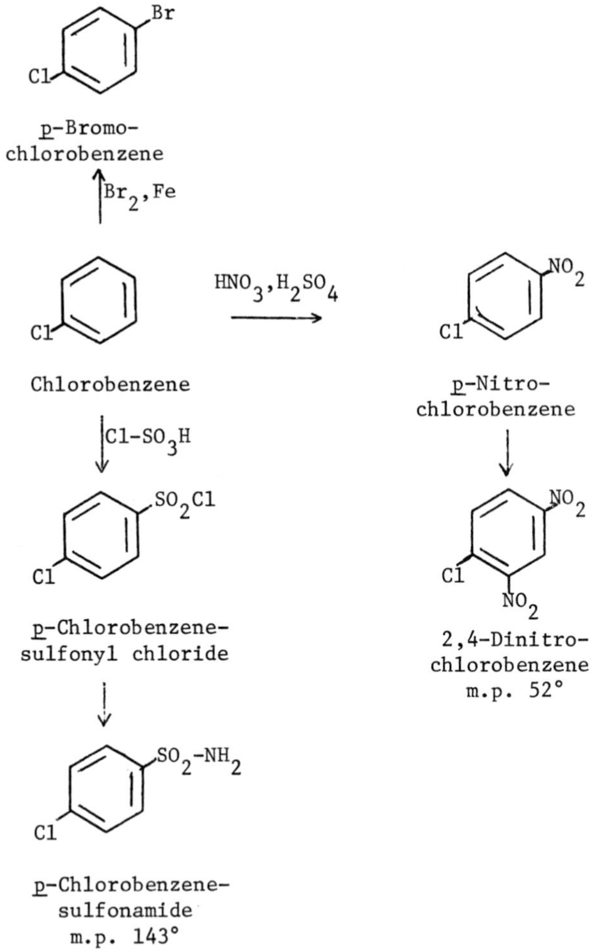

p-Bromo-
chlorobenzene

Br_2, Fe

Chlorobenzene

HNO_3, H_2SO_4

p-Nitro-
chlorobenzene

$Cl-SO_3H$

p-Chlorobenzene-
sulfonyl chloride

2,4-Dinitro-
chlorobenzene
m.p. 52°

p-Chlorobenzene-
sulfonamide
m.p. 143°

Aryl halides having one or more electron-withdrawing substituents ($-NO_2$, $-C\equiv N$, $-\overset{\overset{O}{\parallel}}{C}-R$, $-SO_3H$) positioned ortho or para to the halogen are anomalous and exhibit reactivity to nucleophiles comparable to that of the alkyl halides.

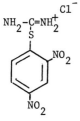

2,4-Dinitroaniline

2,4-Dinitrochlorobenzene   2,4-Dinitrophenol

S-2,4-Dinitrophenyl
isothiouronium chloride

77

EQUIPMENT AND TECHNIQUES

"The cuckoo does not lay its own eggs."

"The hydra descends on its pray and pushes it into its mouth with its testacles."

"The pistol of a flower is its protection against insects."

APPARATUS AND MATERIALS

Every student should own a pair of safety glasses or goggles and should be required to wear these at <u>all</u> times in the laboratory. Accidents involving the eyes are extremely serious and usually arise most unexpectedly and from trivial causes.

Apparatus

Personal Apparatus. The equipment in a student's desk locker or drawer should include:

Six Pyrex test-tubes,	Four Pasteur pipets, 6"
10 x 70 mm	Four Pasteur pipets, 8"
Six Pyrex test-tubes,	Four medicine droppers, 3"
13 x 100 mm	Six rubber bulbs, 2 ml
Three Pyrex test-tubes,	Two microspatulas, 6"
25 x 150 mm	Two glass rods, 3 mm x 180 mm

3. EQUIPMENT AND TECHNIQUES

One test-tube rack, two or
three-row

One test-tube brush

Two small clamps

Three clamp holders

One clamp stand

One iron ring, 3"

One spring-type test-tube
holder

One separatory funnel,
50 or 100 ml

Two filter funnels, 2"

One box filter papers, 9 cm

One beaker, 10 ml

Two beakers, 25 ml

One beaker, 50 ml

Two beakers, 100 ml

One boiling-point tube
(Figure 7a)

One microburner

Two polyethylene wash
bottles, 200 ml

One thermometer, solid stem,
-10 to 360°

Red and blue litmus paper

One suction filtration assembly
(Figure 4b) comprising 1"
Hirsch funnel, rubber (Neo-
prene) adapter and 1" x 7"
Hirsch tube

One graduated cylinder, 10 ml

One conical (Erlenmeyer)
flask, 25 ml

Two conical (Erlenmeyer)
flasks, 50 ml

One conical (Erlenmeyer)
flask, 125 ml

Melting-point capillaries (25)
or microscope cover slips
(50) (for oil bath, Mel-Temp
or Fisher-Johns m.p. apparatus, respectively)

Two watch glasses, about 70 mm

Six soft-glass test-tubes,
10 x 70 mm

Additional Personal Equipment. The student will find it
worthwhile to obtain:

A pair of tweezers	A roll of paper towels
A few razor blades, or an	A roll of toilet tissue
Exacto knife or dis-	Several spring-type clothes pins
secting scalpel	A small roll of absorbent cotton
A few feet of stout metal	A plastic bowl (10" x 8" x 4")
wire (copper or iron)	A small (3-4") aluminum saucepan
A small magnifying glass or	filled with hardened vege-
lens	table shortening (Crisco,
A pair of scissors	Snowdrift, Trex)
An electric "coffee cup	Liquid detergent (Lux, Ivory)
heater" coil	A sponge

Laboratory Equipment. The following equipment should be
at hand in the laboratory or available immediately upon re-
quest. Quantities will suffice for a class of 15-20 students.

Four centrifuges, capacity	Sundry soft glass tubing (2,
six or eight 10 x 70 or	3, 5, 7 and 10 mm
13 x 100 mm test-tubes	Rubber and Tygon tubing (1/8,
Six small (25 and 50 ml)	3/16 and 1/4")
standard-taper round-	Sundry corks
bottomed flasks (Figure	One set of cork borers
2c)	Filter paper (1 cm) for Hirsch
Four standard-taper stillheads	funnels

3. EQUIPMENT AND TECHNIQUES

Six (6") standard-taper Liebig condensers (Figure 2c)	Four burets (10 or 25 ml)
	Four volumetric pipets (10 ml)
	Four volumetric pipets (5 ml)
Four standard-taper thermometers	Soft glass rod (3 and 5 mm)
	Two rough balances, accurate to $\pm$ 50 mg

Variable transformers (Variac) or other power control devices (ten) are desirable, especially if students use "coffee cup" immersion heater coils to heat oil or water baths. Two or more heater coils can be actuated by one control by means of suitable multiple outlet adapters.

Laboratory Instruments

One analytical balance (Mettler H3 or equivalent) accurate to $\pm$ 1 mg or better.

One refractometer (Bausch and Lomb, Hilger, or equivalent).

One oxygen-gas blowtorch. An air-gas blowtorch is useful if a cylinder of oxygen is not available.

Melting-point apparatus (Fisher-Johns or Mel-Temp). If oil baths are used routinely for m.p. determinations, one electrical device, for high melting samples, is sufficient. If electrical devices are to be used routinely, five will be needed.

Solutions and Reagents

Bench-top Reagents and Solutions. The following reagent solutions (100-250 ml each) should be immediately at hand:

Concentrated sulfuric acid	Dilute sulfuric acid (5%, 1N)
Concentrated nitric acid	Dilute nitric acid (5%, 0.8N)
(70%)	Dilute hydrochloric acid
Concentrated hydrochloric	(5%, 1.4N)
acid (36%)	Dilute acetic acid (5%, 0.8N)
Glacial acetic acid	Dilute ammonia (5%, 3N)
Concentrated ammonia (s.g.	Sodium hydroxide solution
0.88, 30%)	(5%, 1.25N)
Sodium hydroxide solution	
(20%, 5N)	

Side-shelf Solutions. The following solutions should be available at each of two side-shelf stations in the laboratory.

Bromine in carbon tetrachloride, 1% v/v.

Congo red solution.

2,4-Dinitrophenylhydrazine solution; 20 g in 500 ml methanol and 500 ml aqueous sulfuric acid (10% v/v).

Ferric chloride in water, 5%.

Ferric chloride (anhydrous) in chloroform, 1%.

Lead acetate in water, 1%.

Phenolphthalein solution; 5 g in 500 ml of ethanol and 500 ml of water.

3. EQUIPMENT AND TECHNIQUES

Phosphoric acid in water, 5%.

Potassium hydroxide in water, 5%.

Potassium permanganate in water, 1%.

Pyridine in chloroform, 1 ml/litre.

Silver nitrate in ethanol, 1% (brown bottle).

Silver nitrate in water, 5% (brown bottle).

Sodium carbonate in water, 5%.

Solutions for Volumetric Analysis. Standard (0.1 N and 1.0 N) solutions of hydrochloric acid and sodium hydroxide should be available when needed, as for determination of the neutralization equivalent of a carboxylic acid or the saponification equivalent of an ester.

Solvents. The common solvents, which should be available in the laboratory include:

Acetone	Ethanol
Benzene	Ethyl acetate
Carbon tetrachloride	Methanol
Chloroform	2-Propanol
Diethyl ether	Hexane or ligroin

Solid and Liquid Reagents. One-hundred-gram quantities of the following solid and liquid reagents, used mainly in the preparation of derivatives should be available at a side-shelf in the laboratory:

Acetic anhydride

Acetyl chloride

Aniline

Benzaldehyde

Benzenesulfonyl chloride

Benzoyl chloride

S-Benzylisothiouronium chloride

p-Bromophenacyl bromide

Chloroacetic acid

Chlorosulfonic acid

Diethylene glycol

Dimethylcyclohexanedione
    (dimedone)

3,5-Dinitrobenzoic acid

2,4-Dinitrochlorobenzene

3,5-Dinitrobenzoyl chloride

2,4-Dinitrophenylhydrazine

Ferrous sulfate

Formic acid

Hydroxylamine hydrochloride

Iodine

Medicinal paraffin

Methyl iodide

2-Naphthol

α-Naphthylamine

p-Nitrobenzoyl chloride

p-Nitrobenzyl bromide

p-Nitrophenylhydrazine hydro-
    chloride

3-Nitrophthalic anhydride

Phenol

Phenylhydrazine hydrochloride

p-Phenylphenacyl bromide

Phosphorus pentachloride

Phosphorus trichloride

Phthalic anhydride

Picric acid

Picryl chloride

Potassium hydroxide (solid
    pellets)

Pyridine

Semicarbazide hydrochloride

Sodium metal (cut into 1/8"
    strips) under oil

Sodium hypochlorite solution
    (Laundry bleach)

Sodium nitrite

Sodium nitroprusside

Styphnic acid (2,4,6-
    trinitroresorcinol)

Thionyl chloride

Thiourea

3. EQUIPMENT AND TECHNIQUES

Toluene-p-sulfonic acid   1,3,5-Trinitrobenzene

Toluene-p-sulfonyl chloride   2,4,6-Trinitrotoluene

p-Toluidine

ORGANIZING THE WORKING SURFACE

"Things" seem to possess an inherent tendency towards
disorder. A working surface will become utterly disorganized
in short order unless attention is deliberately paid to keeping it tidy. Disorder is quite incompatible with semimicro
work.

Keeping order, however, rapidly becomes habit if early
methodical attention is paid to it; and this habit, once
formed, is just as strong as is the habit of disorder.

The following suggestions will help the novice to use his
thirty-inch length of bench to the full:

1. Do not scatter apparatus and materials randomly about
the working surface. Have a specific place for each item.
Move it from this place temporarily, as needed for actual use
and return it immediately afterward.

2. Place all vertical assemblies at the extreme right rear
of the work area. Several vertical assemblies (boiling-point
tube; oil bath in iron ring; Thiele tube; reflux tube or
flask-condenser; Hirsch funnel assembly; separatory funnel;

back-up trap connected to aspirator) may be clamped to one stand. Rarely is an additional stand needed.

3. Place heating devices (electric hotplate; surface-level oil bath; variable transformer) immediately in front of the vertical-assembly "christmas tree."

4. Place a twelve-inch square of paper towel (two thicknesses) immediately to the left of the vertical assembly and at the back of the work area. Routinely, place all clean idle glassware, spatulas and other minor equipment (including supernumerary test-tubes and Pasteur pipets) on this square. Have no idle equipment elsewhere. If you have more apparatus than will fit on the paper square, you have too much clean apparatus out!

5. Place the (two-row) test-tube rack in front of the paper towel square. Place a folded paper towel in the bottom of the back row of the rack and keep clean test-tubes and pipets (without rubber bulbs) upside down in this row, resting on the paper. Keep the front row for tubes actually containing test mixtures.

6. Have three 100-ml beakers containing, respectively, water, methanol and acetone at the center-rear of the work area. Clean each pipet and test-tube immediately after use with an appropriate solvent and (unless it is exceedingly foul) return it to the rack. If water is used, rinse finally with acetone; the item will be dry when it is next needed.

87

3. EQUIPMENT AND TECHNIQUES

7. Have a plastic bowl (approximately 10" x 8" x 4") half-filled with water containing a dash of liquid detergent at the left-rear of the area for equipment which needs washing-up. Rinse out each item as fully as possible (with water and/or solvents) and leave it soaking in the soapy water. When the time comes to carry the bowlful to the sink for cleaning, a quick scrub and rinse will invariably suffice.

8. The remaining area (approximately 12" x 16" at left-front) is quite adequate for the manual operations immediately in hand provided that it is kept uncluttered. Other units (vertical assembly stand; test-tube rack; hotplate) may be moved into this space temporarily, and should be returned as soon as possible.

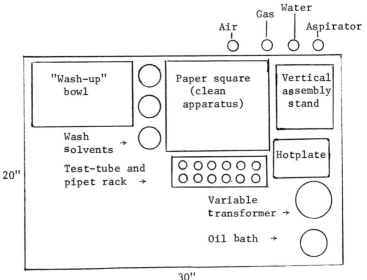

Figure 1. A birds-eye view of a well-organized work area

REACTION, ISOLATION AND PURIFICATION TECHNIQUES

Simplicity is next to godliness. The simpler a piece of apparatus, provided that it will suffice, the better. The humble test tube is unmatched in simplicity, versatility and convenience; few, indeed, are the organic reactions and preparations which can not be carried out perfectly well in test-tubes.

The unit processes commonly used in the identification of organic compounds are essentially the same as those used in preparative work. The most important ones are reaction, isolation, crystallization, determination of physical properties (melting point, boiling point, etc.), together with occasional distillation and solvent-extraction. However, because only small quantities (50 mg to 1 g) of material usually are available, modified ("semi-micro") techniques and apparatus' are convenient. Surprisingly, perhaps, special equipment for small scale work is simple, convenient, and quite inexpensive. The essential requirements are forethought, scrupulous cleanliness and meticulous care. Solid substances are much easier to work with on a small scale than are liquids. As little as 50 mg of a crystalline organic substance is a fair bulk of material whereas even one gram of a liquid is relatively little, and 50 mg barely wets the walls of the container. This is one reason for the extreme emphasis on solid derivatives and on reactions which commonly produce solids rather than liquids.

89

3. EQUIPMENT AND TECHNIQUES

A major consideration in small-scale work is <u>transfer loss</u>.
Clearly, when one is working with 10 g of a solid, the loss of
10-20 mg which may adhere to the Büchner funnel or container is
a minor matter, but with 50 mg or less, such loss is serious
and, if repeated, is catastrophic. And with a liquid to be
distilled, matters are even worse. The key consideration in
the design of semi-micro techniques, then, is avoidance of
transfer losses.

Reaction

Reactions commonly are effected under one of the following
three sets of conditions:

1. <u>Reaction in homogeneous solution at, or near to, room
temperature</u> is most conveniently carried out in test-tubes
(13 x 100 mm). The product may be separated directly from the
reaction mixture by centrifuging if it is a solid or by means
of a Pasteur pipet if it is liquid.

Examples: Anilides and <u>p</u>-toluidides from acid chlorides;
S-benzyl isothiouronium salts from carboxylic acids; dinitro-
phenylhydrazones, semicarbazones, phenylhydrazones and dimethones from aldehydes and ketones; picrates from amines, benzenoid hydrocarbons; chlorosulfonation, nitration and bromination of some reactive aromatic compounds.

2. <u>Heterogeneous reaction at,or near to,room temperature</u> also
is conveniently carried out in test-tubes (13 x 100 mm). The

tube is closed with a soft, well-fitting cork (softened by squeezing its small end thoroughly, or by rolling it on the bench under a hard flat object) or by means of a rubber or plastic stopper. The reaction mixture usually comprises an insoluble liquid or low-melting solid reagent or reactant suspended in an aqueous phase. Very vigorous agitation is essential to ensure the greatest possible surface contact between the phases; hold the tube in the clenched fist between the four fingers and the palm of the hand, hold the stopper in tightly with the thumb, and shake the mixture as violently and continuously as possible.

Examples: Acylation and sulfonylation of phenols, amines and ammonia using the Schotten-Baumann procedure; semicarbazones and phenylhydrazones from water-insoluble aldehydes and ketones.

3. Reaction under reflux. The conventional apparatus (Figure 2c), comprising a small (10-25 ml) standard-taper flask and short (6") Liebig condenser, is merely a scaled-down version of that used in large-scale work. Support the assembly by one clamp which grips the condenser and suspend the flask from the clamp or from a side-arm of the condenser by means of a lightly stretched coil spring or rubber band and a short length of wire twisted about the neck of the flask.

3. EQUIPMENT AND TECHNIQUES

A perfectly satisfactory and extremely inexpensive alternative for almost all semimicro preparative work is simply a long (7-8"), relatively narrow (15-20 mm internal diameter) tube (Figure 2a). Confine the heating to that portion of the tube which actually contains the mixture and control it so as to ensure gentle reflux rather than vigorous boiling. Plug the mouth of the tube with a slit cork or some absorbent cotton. The walls of the long tube provide adequate reflux condensation of the common solvents, including pyridine, acetic acid, formic acid, water, water-alcohol mixtures, 2-propanol, ethanol, and even methanol. Occasional additions of fresh solvent may be required if the reaction time is unduly long. Even the more volatile solvents and reactants such as thionyl chloride carbon tetrachloride, chloroform, the lower alkyl halides, and even ether and carbon disulfide can be contained satisfactorily (subject to occasional topping-up) if the tube is fitted with a simple U-tube cold finger (Figure 2b) which the student can make in a few minutes.

Heat the reaction mixture by means of a hot bath (Figure 3), or, for very high-boiling mixtures (above 200°) with a microburner flame directly. Heat only that portion of the flask or tube which actually contains the mixture. "Bumping" (violent ebullition) may be minimized by adding a very small (pinhead size) portion of porous tile or proprietary boiling stone ("Boileezer"); bumping is rarely a problem with the test-tube device since the upper part of the tube is big enough to contain the whole volume of liquid.

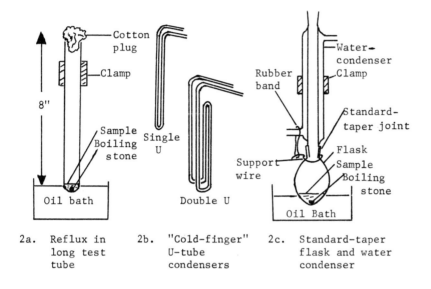

2a. Reflux in long test tube

2b. "Cold-finger" U-tube condensers

2c. Standard-taper flask and water condenser

Figure 2. Assemblies for reaction under reflux

To make a single-U "cold-finger" condenser (Figure 2b), heat the center of a ten-inch length of 3 mm outside-diameter glass tubing in a <u>small</u> microburner or blowtorch flame until it is soft. Bend the tube into a <u>narrow</u> U which can be inserted into the long reflux test-tube (Figure 2a). A slight kink in the bend will not matter as long as it does not unduly restrict the flow of water. Heat both arms of the U about one inch from the open ends and bend them over as shown in Figure 2b.

3. EQUIPMENT AND TECHNIQUES

Fire-polish the open ends of the U-tube.

Connect one of the open ends through a short length of narrow (1/8") "Tygon" or rubber tubing to the narrow end of a medicine dropper or short Pasteur pipet and connect the wide end of this to the water faucet with 1/4" rubber hose. Connect the other open end of the condenser to the drain. Insert the U into the long test-tube. Plug the mouth of the tube around the condenser outlet-tubes with cotton.

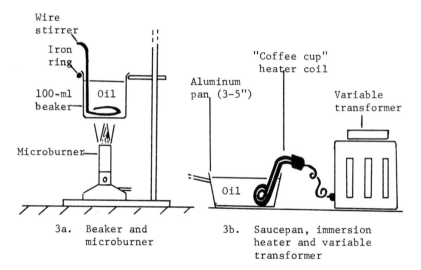

3a. Beaker and microburner

3b. Saucepan, immersion heater and variable transformer

Figure 3. Heating baths

The most generally satisfactory heat-transfer medium is hardened vegetable oil or shortening ("Crisco," "Snowdrift," "Trex"). It melts readily and does not smoke or constitute a fire hazard even at 200°C. Additionally, it sets solid on cooling so that the risk of messy spills in a desk drawer are

much reduced. Mineral oil (medicinal liquid paraffin,"Nujol")
is much less satisfactory. Water is useful for some purposes.

Heating sources are interchangeable, of course; a saucepan
can be heated with a microburner, a beaker with an immersion
heater, or either with an electric hotplate. With an immersion
heater the bath can be placed directly on the working surface,
a distinct advantage.

Transfer of Mixture After Reaction Under Reflux

When reaction under reflux is complete, the mixture must
be transferred to another vessel (usually one or more 13 mm x
100 mm centrifuge tubes) without incurring undue transfer loss.
Satisfactory procedures are:

1. If the volume of the mixture is greater than about 10 ml,
stir up any solid thoroughly, free it from the walls of the
container with a spatula if necessary and pour the suspension
from the reaction vessel. Allow ample drainage time,especially
if the reaction vessel is a long tube, by clamping it, inverted
over the receiver, for several minutes. Wash out the reaction
vessel with 1-2 ml of reaction solvent or a suitable wash
liquid or diluent. Ensure that all the inner walls are
thoroughly rinsed and any residual solid taken into suspension.
Drain the washings into the receiver as before.

2. If the volume of the reaction mixture is small (below 5-
10 ml), the product is a solid-liquid mixture, and the solid

95

will not melt or dissolve in the liquid even upon heating, add
more solvent, bring the solid into suspension, and transfer the
mixture as in (1) above.

3. If the volume of the reaction mixture is small (below 5 ml)
and it is completely liquid or can be liquified by heating,
withdraw it completely into a Pasteur pipet (preheated if necessary; an 8" pipet is required if the mixture is in a long reaction tube) and extrude it into the receiver. Wash out the reaction vessel with a small volume (10-50% of the volume of the
reaction mixture) of fresh solvent (hot, if necessary). Draw
this up into the pipet several times and extrude it into the
reaction vessel in such a manner that all of the inside walls
of the vessel and the pipet are very thoroughly washed with
this small amount of solvent. Transfer the wash liquid to the
receiver.

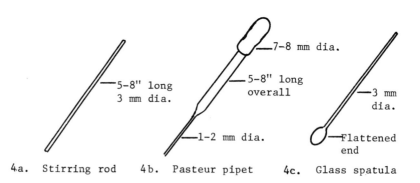

4a. Stirring rod   4b. Pasteur pipet   4c. Glass spatula

Figure 4. Transfer implements

REACTION, ISOLATION AND PURIFICATION TECHNIQUES

To make two Pasteur pipets, heat the middle of a 10-15 inch length of 7-8 mm diameter tubing over a Meker-burner flame, rotating the tube between the fingers and thumbs so that the mid-section is uniformly softened. Remove the tube from the flame and gently pull the ends apart until the hot section is long (4-6") and narrow (1-2 mm). When the tube has cooled, scratch the middle of the narrow portion transversely with a file, glass cutter or diamond. Slight stress will then break it at this point. The cut capillary ends should be fire-polished and strengthened by touching them to a flame for an instant.

To make a glass spatula, heat one end of a glass rod (5-8" x 3 mm) until it is soft, and squash the end between two asbestos boards. Anneal the squashed end by heating until it is almost soft, and allow it to cool slowly.

The ends of a stirring rod (5-8" x 3 mm) should be fire-polished by heating them until they become soft and smoothly rounded.

Separation and Washing

A reaction used to produce a solid derivative typically produces the solid suspended in a liquid reaction mixture. If the mixture is not in a test-tube which fits the centrifuge, transfer it as completely as possible to one or more such tubes. Spin the tube(s) in a centrifuge until the solid (or almost all

3. EQUIPMENT AND TECHNIQUES

of it) separates as a compact mat below the clear supernatant liquid. If necessary (as with fine needle-like crystals), crush the mat with a thin glass rod (3 mm diameter, 10 cm long) and spin again. Pipet off most of the supernatant liquid and drain out the remainder by carefully inverting each tube and allowing it to stand on its mouth on two or three thicknesses of filter paper for several minutes (Figure 5a). The mat should adhere to the bottom of the tube. If it falls out, lift it from the paper with a spatula and return it to the tube. To wash the product, add the wash liquid (usually water), stir the solid thoroughly, centrifuge, and remove and drain away the supernate as before.

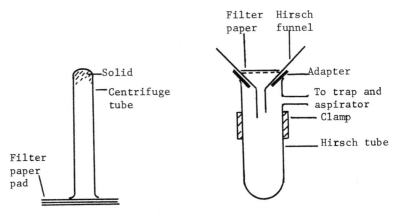

5a. Simple draining

5b. Hirsch funnel assembly for filtration with suction

Figure 5. Assemblies for final separation of solid from mother liquor

REACTION, ISOLATION AND PURIFICATION TECHNIQUES

Crystallization

Purification of a solid substance by crystallization is conveniently effected by preparing a saturated or near-saturated solution in a hot solvent and cooling it; most substances are less soluble in cold than in hot solvents. As long as the temperature at which the solute begins to separate is significantly (typically 20° or more) below its melting point, it will usually separate as a solid. Impurities will remain dissolved as long as the solution is not saturated with them. For significant purification to be achieved, solid must separate directly from the solution as it cools. A solid produced by cooling material which separated initially as a second liquid phase (an "oil") will be little or no purer than the starting material; neither will a precipitate produced by adding a weak solvent to a solution in a more powerful solvent, nor a solid produced by evaporation of a solution.

A second crystallization usually is necessary to ensure purity, and several successive recrystallizations may be required. Successive crystallizations invariably are achieved more readily than was the first, especially if the solutions are "seeded" with crystals of the substance retained from a previous crystallization. A product should be recrystallized until the melting point no longer is changed by further recrystallization. Two crystallizations from solution usually are adequate.

3. EQUIPMENT AND TECHNIQUES

After each crystallization, centrifuge the mixture and remove most of the supernatant liquid by means of a pipet. Invert the tube upon filter paper and allow the remaining liquid to drain away before adding fresh solvent for recrystallization. No transfer loss whatsoever will be incurred since the material remains in one vessel throughout any number of recrystallizations.

The solvents from which organic solids are commonly crystallized, listed roughly in order of decreasing use, are:

Ethanol (95% or absolute), methanol, 2-propanol, water, acetone, ethyl acetate, glacial acetic acid, chloroform, benzene, carbon tetrachloride, petroleum ether (light petroleum; ligroin) and diethyl ether.

Mixtures of two or more of the above solvents are often useful. The alcohols, alone or mixed with water, usually offer the best chance of success and are tried first unless there is some specific reason to prefer another solvent. A major advantage to the use of water-miscible solvents (alcohols, acetone, acetic acid) is that the crude product, which usually is wet with water, need not be dried completely before the solvent is added.

Crystallization is an art, mastered only through experience. The conditions required vary with the substance so that no completely general procedures can be given. One or the other of the techniques outlined below usually will give good results.

REACTION, ISOLATION AND PURIFICATION TECHNIQUES

1. <u>Crystallization</u> <u>from</u> <u>a</u> <u>single</u> <u>solvent</u>. Add to the solid mat of crude product in the test tube a <u>small</u> amount of the chosen solvent. Heat the mixture to boiling (to 100° in a boiling water bath for solvents of b.p. above 100°). Shake it, hot, for several minutes. If the solid does not dissolve (this includes cases where it melts to form an oil, but does not dissolve) add more solvent, portionwise, and heat and shake the mixture until it does. Frequently a small proportion of the crude sample will not dissolve. In such cases, do <u>not</u> continue to add solvent, but pipet the clear solution into a clean tube and discard the insoluble residue.

Cool the solution. In some cases rapid cooling, with or without agitation, will cause satisfactory crystallization but in others slow cooling is better. If the substance separates as an oil, reheat the liquid until the oil is redissolved and cool the solution while rubbing the inside wall of the tube at and below the surface of the liquid with a glass rod. If the product persists in separating as an oil, reheat the solution and dilute it with a little more solvent so that, upon cooling, separation commences only at a rather lower temperature. In difficult cases of oiling-out, dilute the solution to such an extent that separation (turbidity) barely commences at room temperature. Rub the inner wall of the tube well and cool the solution in ice or in a refrigerator.

3. EQUIPMENT AND TECHNIQUES

In some cases a supersaturated solution is obtained and crystallization does not occur even on cooling to room temperature or below. Such a solution should be "seeded" with the crude solid substance. On the other hand, the substance may be so soluble that a saturated solution is not obtained even with the smallest amount of solvent which can be used conveniently. Such a solution should be diluted with a "weaker" solvent and treated according to procedure 2 (mixed solvents) below.

2. Crystallization from mixed solvents. The solvents to be used in admixture must, of course, be miscible with each other. Dissolve the solid in the minimum amount of the more powerful solvent (ethanol or methanol, for example) at a temperature somewhat below the boiling point of either of the solvents to be used. Add to the hot solution the specified proportion of the weaker solvent (water, for example) or, if no proportion is specified, an amount sufficient to make the solution slightly turbid. If the volume of the resulting solution is inconveniently small, add more of the stronger and weaker solvents in appropriate amounts. Cool the solution to induce crystallization. Use the methods described above to overcome problems arising from supersaturation and separation as an oil.

REACTION, ISOLATION AND PURIFICATION TECHNIQUES

Isolation and Drying

After the final recrystallization, wash the product with a little cold solvent or solvent-mixture, centrifuge the mixture, and drain the wash liquid away thoroughly. If well-formed crystals have been obtained, a gentle tap on the inverted tube will usually dislodge the mat of crystals. Allow it to fall onto a clean dry filter paper, upon which it will dry completely in the air in a few minutes. Occasionally, waxy or leafy crystals result which are not readily dislodged, or which hold so much wash liquid that drying is tedious. In such cases, suspend the solid in a small amount of cold solvent or solvent mixture (preferably a somewhat "weaker" mixture than that used for recrystallization) and filter the mixture, with suction, using a Hirsch funnel assembly (Figure 5b). Suck the solid dry and carefully lift the paper bearing it out of the funnel onto a large filter paper and allow it to dry completely in the air. The transfer loss incurred in this filtration is tolerable since it occurs only once in the purification procedure.

Removal of Solid, Tarry and Colored Impurities before Crystallization

A crude reaction product often is contaminated with insoluble materials, tars, and colored and colloidal materials which must be removed before recrystallization is attempted.

3. EQUIPMENT AND TECHNIQUES

Prepare a filter pipet by plugging the nozzle of a long medicine dropper or short-nosed (broken) Pasteur pipet with absorbent cotton. The cotton is best inserted through the wide end of the pipet and pushed into the narrow end with a stout wire or a thin glass rod. The plug must be packed tightly enough to withhold solid impurities, but not so tightly that liquid will not flow through it readily. The cotton should protrude through the narrow end of the pipet so that the plug may be gripped with tweezers and pulled out.

Figure 6.

Nozzle of filter pipet

Dissolve the sample in a suitable amount of hot solvent or solvent-mixture exactly as if it were to be crystallized directly. The volume of liquid should be such that it can be completely drawn into the pipet. Place the filter pipet in the tube and boil the solution gently for a few minutes so that the pipet is thoroughly heated by the vapors of the boiling solution. Lift the hot pipet just out of the liquid, squeeze the air out of the rubber bulb, and draw the liquid up through the cotton plug. Remove the pipet from the tube, quickly pull out the cotton plug and discharge the hot solution into a clean tube.

104

REACTION, ISOLATION AND PURIFICATION TECHNIQUES

Caution: The pipet must be thoroughly preheated and the transfer operation must be effected rapidly; otherwise crystallization may occur in the pipet.

Colloidal, colored and tarry impurities often are not adequately removed by filtration through cotton. These may be removed by boiling with decolorizing carbon. Dissolve the sample in a slight excess of the selected recrystallization solvent, cool the solution somewhat, add a small amount (typically one-twentieth of the bulk of the solution) of decolorizing carbon ("Norit," "Darco") and boil the mixture gently for a few minutes. Separate the hot solution from the carbon by means of a filter pipet. It is usually desirable to wipe the outside of the pipet with a piece of toilet tissue before removing the cotton plug and discharging the solution. A second filtration into a clean filter pipet may be necessary to remove all traces of carbon from the solution.

Distillation

This technique is relatively little used in elementary semi-micro work; in fact, it is deliberately avoided wherever possible since distillation of small volumes (0.5 to 1 g) of sample leads to considerable losses on the relatively large wetted areas of even the smallest apparatus. Volatile solvents (reaction or extraction solvents) in the volumes usually encountered (10-20 ml) are conveniently removed by simple

3. EQUIPMENT AND TECHNIQUES

evaporation under reduced pressure and are not recovered.

Occasionally, badly discolored samples of such substances as phenols and amines may merit purification by distillation. A useful device is a semi-micro version of the classical distillation retort (Figure 7a) which is readily made from a test tube. It is filled with the liquid (molten) sample by means of a long-nosed Pasteur pipet inserted into the side arm of the retort. During the distillation the side arm should be inserted well into the receiver tube which may be cooled in ice if the distillate is volatile (b.p. below 80-100°).

An alternative to the retort is a small side-arm test tube (Figure 7b). Assemblies involving standard-taper joints and water condensers are to be avoided with samples smaller than 3 ml since they lead to unacceptable wetted-surface losses.

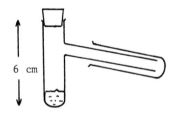

6 cm

7a. Microretort and
     receiver

7b. Side-arm test tube
     and receiver

Figure 7. Apparatus for distillation

REACTION, ISOLATION AND PURIFICATION TECHNIQUES

To make a microretort, roll a sheet of asbestos paper
(4" x 4") tightly about a small(13 x 100 mm) test tube. Pull
the test tube out until only about 1 cm at its closed end is
still encased in the asbestos and wire the asbestos tightly
to the end of the tube. Hold the asbestos in one hand, the
open end of the tube in the other, heat the glass strongly just
where it comes out of the asbestos, rotating it to attain uni-
form heating, and,when it is soft, draw it out several inches
so that the diameter of the constricted part is about 3 or 4
mm. Reheat the glass close to the asbestos and bend it over
rather more than 90°. Cut the constricted tube two or three
inches from the bend and fire-polish the cut end.

Extraction with Solvents

Certain reactions produce water-solutions of a desired
organic product together with other materials (inorganic salts)
in such concentrations that the desired product cannot be recovered by evaporation. A typical example is the oxidation of
an alkylbenzene to a benzenoid carboxylic acid using potassium
permanganate solution. If, as is often the case, the product
is soluble also in a water-immiscible solvent, it may be ex-
tracted by agitating the aqueous layer with that solvent and
recovered by evaporating the solvent phase. The most useful
solvents are diethyl ether, dichloromethane, chloroform, benzene, and petroleum ether (ligroin).

3. EQUIPMENT AND TECHNIQUES

If the volume of the aqueous reaction mixture is 20 ml or greater, transfer it to a 50 ml separatory funnel (Figure 8)

Figure 8.

A separatory funnel

and extract it with several successive 10 ml portions of the solvent. If the solvent is dichloromethane, chloroform or carbon tetrachloride, it will form a lower layer which may be run off directly into a receiver and the next portion of solvent added directly to the aqueous layer in the funnel. If the solvent is less dense than water (ether, benzene, ligroin), run off the lower aqueous layer into a suitable container, drain the solvent layer into a receiver, and return the aqueous layer to the funnel. Repeat the process as desired.

If the volume of the aqueous layer is relatively small (<20 ml), carry out the extraction in a test-tube. Use small volumes (1 ml or one quarter of the volume of the aqueous layer, whichever is larger) of solvent and separate the solvent phase by means of a Pasteur pipet.

Whether a separatory funnel or a test-tube is used, stopper the vessel and shake it gently, holding in both stopper and stopcock firmly. Remove the stopper briefly to release excess pressure, replace it and shake the mixture vigorously to ensure very thorough phase contact. Hold the stopper in securely during each shaking else it may be blown out and the mixture lost. Allow the two phases to separate completely before removing either of them.

Drying Solutions in Organic Solvents

The combined solvent phases resulting from extraction of an aqueous solution will contain small amounts of water. This must be removed before the solution is evaporated, a purpose usually accomplished by shaking the solution with a solid substance (drying agent) which will combine with the water but not with the solvent or the desired solute.

Four suitable drying agents are listed below:

Drying agent	Advantages	Disadvantages
Anhydrous $MgSO_4$	Rapid; high water capacity.	May react with some alcohols and amines.
Anhydrous Na_2SO_4	Virtually inert chemically.	Slow in action.
Anhydrous $CaSO_4$ (anhydrite)	Extremely powerful.	Reacts with alcohols and amines; low water capacity.
Anhydrous K_2CO_3	Inert to amines and alcohols; fairly rapid.	Reacts with acids.

To dehydrate the solvent phase, add to it one-twentieth of its bulk of the selected desiccant and shake the mixture frequently during 10-20 minutes ($MgSO_4$) or allow it to stand with occasional shaking for two hours or more (Na_2SO_4). Separate the dried solution from the spent desiccant either by filtration with suction (Figure 5b) or by means of a filter pipet (Figure 6).

Evaporation of Solutions in Solvents

a) Place the dried solution in a conical flask fitted with a well-fitting rubber stopper and outlet tube (Figure 9).

Figure 9.

Evaporation under
reduced pressure

The flask should be of such size that it is less than one-third filled. Connect the outlet tube to a water aspirator via a "trap," immerse the flask in warm or hot (depending on the b.p. of the solvent) water and agitate it continuously until the solvent is removed.

b) Place the solution in a beaker of such size that it is less than one-third full. Place the beaker on a warm hot-plate or in a hot water bath, and direct a gentle stream of compressed air towards the surface of the liquid.

Water from an aspirator often "sucks back." The vessel containing the desired material should be connected to a large trap (a Büchner flask is suitable) which, in turn, is connected to the aspirator. All connections in the assembly, must of course, be reasonably vacuum-tight.

DETERMINATION OF PHYSICAL PROPERTIES

The chemical properties of an organic substance identify the nature of the functional groups attached to the carbon backbone and allow its designation as one of a particular class of compounds. The physical properties of the substance and of substances derived from it by means of predictable reactions of the functional groups permit its identification as one member of that class. The most useful and most conveniently determined physical properties are the melting point (solids), the refractive index and the relative density (liquids), and the boiling point (liquids and some solids).

Determination of Melting Point

The melting point of a pure solid substance is a reliable, reproducible physical characteristic which may be determined easily and accurately using simple apparatus. This is a major reason for the practice of converting unknown substances into solid derivatives in order to identify them. A melting range (0.5 to 2°), rather than a precise melting point is usually observed and should be recorded.

Approximate melting-point determination (Figure 10). Support a metal plate horizontally above a bunsen burner or microburner. Fix a thermometer horizontally with its bulb about 1" above the center of the plate and place a minute portion of sample on the upper side of the bulb. Heat the metal plate.

111

3. EQUIPMENT AND TECHNIQUES

The hot air rising from the plate heats the thermometer bulb and the sample. Note the temperature at which the sample liquifies on the bulb. Melting points accurate to ±5° may be obtained by this method after some practice.

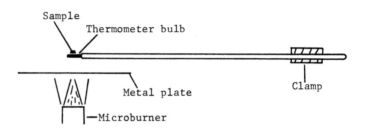

Figure 10. Approximate melting point determination

Accurate melting-point determination. Three types of procedure are commonly used. Two involve commercial electrical devices while the third (which this author prefers) uses a simple oil bath.

1. Electrical hot plate (e.g. Fisher Johns apparatus or Kofler hot-stage microscope, Figure 11a). Crush a minute amount of the solid sample between two very thin (0.1 mm) pieces of glass (microscope cover-slips). Place the sandwich on the electrically heated hot plate of the apparatus, cover it, and observe it through the lens system or microscope. Raise the temperature rapidly to a value some 10-20° below the melting point (as determined roughly, or expected) and then slowly (2-3° per minute) until melting occurs. Read the temperature

from the thermometer, the bulb of which is embedded in, and
makes good thermal contact with the hot plate.

2. Electrical air bath (e.g. Mel-Temp apparatus, Figure 11b).
Seal one end of a two-inch length of narrow (~1 mm) thin-walled
glass tubing (melting-point capillary) by touching that end to
a burner flame. Push the open end downwards several times into
a small pile of the solid sample on a hard (glass or porcelain)
surface, invert the tube and tap it so that the sample trapped
in the open end falls into the sealed end. The heating device
consists of a coil of resistance wire approximately one inch
long and half an inch in diameter, closed at both ends, but
having holes so arranged that a thermometer bulb and the sealed
ends of one or more capillaries may be inserted,are illumi-
nated, and may be observed through a lens. Insert the thermometer bulb and the loaded tube and raise the temperature
rapidly to 10-20° below the melting point and then slowly
(2-3° per minute). Observe and record the melting point or
range.

3. Oil bath (Figures 11c, d). Introduce the sample into a
capillary tube as described above. The oil bath may be a small
(50-100 ml) beaker (Figure 11c) or a Thiele tube (Figure 11d)
and should be equipped with a simple stirrer made from a length
of stout wire. Immerse the thermometer bulb and the sealed end
of the loaded tube in the oil, close together, and heat the

113

3. EQUIPMENT AND TECHNIQUES

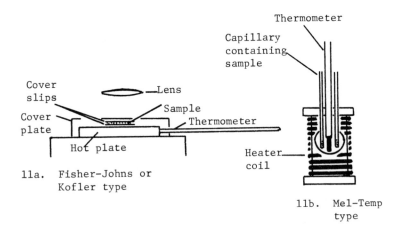

Cover slips

Lens

Cover plate

Sample

Thermometer

Hot plate

11a.   Fisher-Johns or Kofler type

Thermometer

Capillary containing sample

Heater coil

11b.   Mel-Temp type

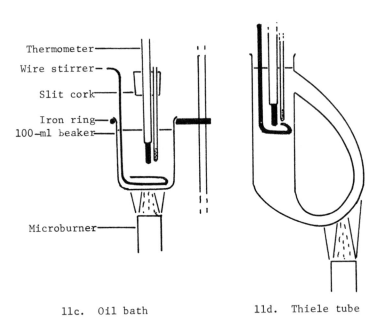

Thermometer

Wire stirrer

Slit cork

Iron ring
100-ml beaker

Microburner

11c.   Oil bath

11d.   Thiele tube

Figure 11.   Assemblies for accurate melting-point determination

bath rapidly to a temperature 10-20° below the melting point of the sample and then slowly (2-3° per minute) with <u>vigorous stirring</u> until melting occurs. The sample may be observed with the naked eye or through a lens.

The capillary tube is best supported by means of a slit cork. Drill a hole longitudinally through a small cork. Insert the thermometer, which should fit the hole snugly, so that the cork is about 1.5 inches above the bulb. Cut a longitudinal slit in the side of the cork and wedge the loaded capillary tube into the slit so that its sealed end is adjacent to the thermometer bulb. An alternative to the slit cork is a short (1/8") length of rubber tubing but this discolors the oil.

Determination of Boiling Point

Accurate determination of boiling points, especially with small samples, is not a simple matter. Approximate values ($\pm$ 5-10°) may be obtained using either of the procedures given below. The first uses somewhat more material but is preferred since it is less likely to be affected by traces of low-boiling impurities. The used sample can be recovered.

1. <u>Reflux</u> <u>procedure</u> (Figure 12a). The boiling-point tube is a long (2" shorter than the thermometer) narrow (inside diameter twice the thickness of the thermometer stem) glass tube sealed at one end. Place a pinhead-sized piece of boiling stone ("Boileezer") and 5-10 drops of the liquid in the tube and support the thermometer axially within the tube by

3. EQUIPMENT AND TECHNIQUES

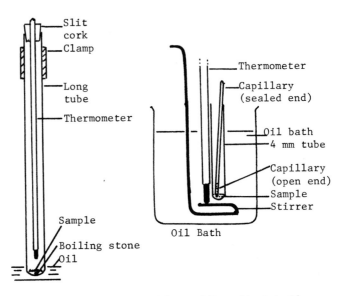

Figure 12a. Reflux apparatus for boiling point determination

Figure 12b. Siwoloboff procedure for boiling-point determination

means of a slit cork so that its bulb is one-half inch above the liquid sample. The bulb must not touch the walls of the tube. Support the assembly vertically and heat the bottom quarter-inch of the tube in a stirred oil bath until the sample boils. Note the temperature when steady reflux equilibrium is obtained, with the thermometer bulb completely immersed in the condensing vapor.

2. <u>Capillary tube (Siwoloboff) method.</u> Prepare a m.p. capillary and a sample tube (a 2" length of 3-4 mm i.d. tube with one end sealed). Place 1-2 drops of liquid in the sample tube and drop the capillary, open end downwards, into this. Immerse

the sample tube in an oil bath. Heat and stir the oil. A slow stream of bubbles will rise from the open mouth of the capillary. At the boiling point of the sample, this stream will increase suddenly and most markedly. Note the temperature. Stir the oil and allow it to cool. When the bath temperature falls to the boiling point of the sample, the flow of bubbles will cease completely. Again note the temperature. To repeat the determination remove the capillary and replace it with a fresh one. This method is reliable as long as traces of low-boiling impurities are absent but requires some practice.

Refractive Index

The refractive index of a liquid is a physical characteristic which can be measured readily with high precision (5 significant figures) using an Abbé refractometer. The parameter measured is the maximum angle at which a beam of light, passing into a film of the liquid from a glass prism of high refractive index, is totally reflected at the glass-liquid interface. This angle is simply related to the refractive index of the liquid. The refractive index varies with the wavelength of the light used and usually is quoted for 5893Å (the sodium D line). Most refractometers incorporate compensating devices so that the refractive index is that for 5893Å even though white light is used.

3. EQUIPMENT AND TECHNIQUES

The instrument (Figure 13) comprises, essentially, a small telescope with a compensating prism in its objective, a split prism, one-half of which is independently movable, and a mirror. To carry out a determination, open the split prism, place two drops of the liquid sample on the horizontal plane-polished face and close the prism so that the liquid completes optical contact between the two halves. Set the telescope at a low refractivity reading (about 1.3), adjust the mirror to illuminate the field of view and focus the eyepiece on the crossed-hairs within the telescope. By means of the adjusting wheel, move the interlocked telescope and split prism until the lower half of the field becomes dark. Rotate the compensator at the objective of the telescope horizontally until the dark-light boundary becomes sharp and colorless and then again rotate the adjusting wheel until the boundary coincides exactly with the intersection of the crossed hairs. Read the refractive index through the lens over the scale beside the telescope.

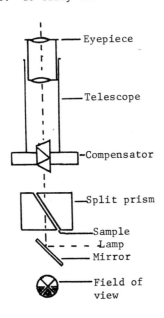

Figure 13.

Simplified diagram of the Abbé refractometer

Clean the split-prism immediately by opening it, and gently wiping off liquid sample, first with absorbent cotton barely moistened with acetone and then with cotton well wetted with the solvent. Leave the prism open until it is dry.

Do not use the instrument with corrosive liquids such as acids, phenols and amines. These are readily identified by other means.

A refractive index is usually quoted for 20 or 25° (n_D^{20}) and should be determined at these temperatures if possible. The instrument provides means for water from a thermostatically controlled bath to be circulated about the prism if necessary. The refractive index of an organic liquid typically decreases by about 0.0004 units for each degree centigrade increase in temperature.

The refractive index is very sensitive to impurities in the sample. The material should be distilled and the determination carried out on a middle cut of the distillate.

Relative Density

1. _Capillary tube method_. Weigh accurately a clean, dry, dust-free melting-point capillary tube, which should be 3-4 inches long and open at both ends. Hold the tube with tweezers and touch one end to distilled water in a small beaker (Figure 14a). Water will rise in the tube and can be induced to fill it completely while the end of the tube is only _just_ touching

the water by tilting the tube almost to a horizontal position.
Quickly withdraw the tube holding it horizontally so that it
remains completely filled. With a toilet tissue, carefully
wipe off any water adhering to the outside of the tube but do
not allow any water to be drawn from the inside or the filling
procedure must be repeated. Immediately weigh the filled tube.

Rinse out the water with several ml of acetone (from a
fine-nosed Pasteur pipet) and then with several ml of diethyl
ether or pentane. Dry it thoroughly by blowing air through it
from a Pasteur pipet.

Fill the tube with the sample exactly as you filled it with
water, again remove any surplus which adheres to the outside,
and weigh the tube filled with the sample. Note and record
the temperature of the laboratory.

2. Pycnometer method. The pycnometer (Figure 14b) is essentially a large-capacity capillary tube. To fill it, attach a
short length of narrow rubber or plastic tubing to end B and
draw liquid in through end A ensuring that no bubbles of air
are entrained. Remove the flexible tube and wipe the ends with
toilet tissue taking care not to draw liquid out of the filled
pycnometer.

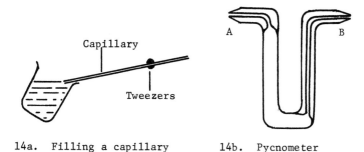

14a. Filling a capillary   14b.   Pycnometer

Figure 14.   Relative density determination.

If W_1 = weight of empty tube or pycnometer

$\quad W_2$ = weight of tube filled with water

$\quad W_3$ = weight of tube filled with sample

Then $\dfrac{W_3 - W_1}{W_2 - W_1}$ is the density of the sample relative to the

density of water at the temperature of the laboratory. Densi-

ties are normally recorded relative to that of water at 4°; for

this the above value should be multiplied by the density of

water at the laboratory temperature (0.997 at 25°).

The third weighing must be executed very rapidly if

the sample is very volatile (lower aliphatic ethers, hydro-

carbons,etc.); otherwise loss by evaporation will lead to

erroneous results.

4

FUNCTIONAL GROUP IDENTIFICATION

"For dog bite: Put the dog away for several days. If he has not recovered, then kill it."

"For snake bite: Bleed the wound and rape the victim in a blanket for shock."

"For fractures" To see if the limb is broken, wiggle it gently back and forth."

The diagnostic and confirmatory tests detailed in this chapter will suffice to differentiate and identify the thirteen common functional-group classes.

Amines	Aldehydes	Alkenes
Carboxylic acids	Ketones	Alkanes
Phenols	Alcohols	Aromatic hydrocarbons
Esters	Ethers	Alkyl halides
		Aryl halides

A tabular summary of the sets of tests occupies the next two pages.

The scheme is flexible. Appropriate interpolation of additional tests, many of which the experienced student can devise for himself, will permit the recognition of amides, epoxides, acid anhydrides, acid halides, nitro compounds, sulfonic acids and alkynes, as well as some polyfunctional compounds.

123

4. FUNCTIONAL GROUP IDENTIFICATION

SUMMARY OF FUNCTIONAL GROUP –

General Physical Properties

1. Obvious properties; color, odor, physical state.

2. Approximate melting point (solids) or boiling point (liquids).

3. Solubility in water; acidity or basicity.

4. Elements present (N, Cl, Br, I) other than C, H and O.

Diagnostic Sequence

1. Solubility in dilute mineral acid; reprecipitation by alkali; diagnosis of organic base.

2. Solubility in dilute caustic alkali; reprecipitation by acid; diagnosis of acidic substance (acid, phenol).

3. Hydrolysis by hot caustic alkali; diagnosis of ester.

4. Precipitate with 2,4-dinitrophenylhydrazine reagent; diagnosis of aldehyde or ketone.

5. Reaction with metallic sodium; diagnosis of lower alcohol.

6. Reaction with cold concentrated sulfuric acid; reprecipitation upon dilution with water; differentiation between alcohol or alkene, ether, and hydrocarbon or halogenated hydrocarbon.

7. Combustion; differentiation between aliphatic and aromatic substance.

4. FUNCTIONAL GROUP IDENTIFICATION

- DIAGNOSIS AND CONFIRMATION

Confirmatory Tests

1. Amines.
   a) Reaction with nitrous acid; differentiation of aliphatic from aromatic and between primary, secondary and tertiary amine.
   b) Hinsberg's test; differentiation between primary, secondary and tertiary amine.

2. Carboxylic acids and phenols.
   a) Liberation of carbon dioxide from sodium bicarbonate; differentiation of carboxylic acids from most phenols.
   b) Solubility in sodium carbonate solution; differentiation of carboxylic acids from most phenols.
   c) Color with aqueous ferric chloride; most phenols.
   d) Color with ferric chloride/pyridine in chloroform. Almost all phenols.

3. Esters. Hydroxamate - ferric chloride color test.

4. Aldehydes and ketones.
   a) Tollens' test; aldehydes.
   b) Iodoform test; methyl ketones.

5. Alcohols. Reaction with acetyl chloride.

6. Alkenes.
   a) Decolorization of bromine in carbon tetrachloride.
   b) Baeyer's test.

7. Ethers. Ferrox color test.

8. Halogen compounds. Mobility of halogen.

4. FUNCTIONAL GROUP IDENTIFICATION

GENERAL CONSIDERATIONS

The crucial importance of reliable functional-group identification cannot be overemphasized. Minutes saved in a
skimped classification can result in hours wasted in attempts
to prepare derivatives of a functional group which never was
present!

Each test takes very little time if you are properly prepared and equipped, and keep your work area well-organized. It
is not a bad idea to carry out each test twice. This is manda-
tory where a positive indication (i.e. an indication of the
presence of a functional group) is obtained or where a large
investment of time and material will be based on the result.

A blank, that is, a test performed under identical conditions using all the reactants except one (usually the unknown)
should be carried out alongside any test, so that the behavior
of the sample can be compared with this reference point. Again,
especially with critically important tests, it is worthwhile
to perform the test on a sample of a known substance containing
the particular functional group the test should detect. For
example, if you are testing for acidity by observing the reaction of a substance with sodium bicarbonate solution, determine simultaneously the effect of a neutral substance (sugar
or ethanol) and a known acid (benzoic or acetic) on separate
portions of the bicarbonate. The reference substances should
be as similar as possible to the unknown. Thus a solid

126

suspected acid might be compared with sugar and benzoic acid;
a liquid with ethanol and acetic acid.

Diagnostic and Confirmatory Tests

The different purposes and applications of diagnostic
tests and confirmatory tests must be very clearly appreciated.

Diagnostic tests comprise a sequence of experiments designed to determine the absence and suggest the presence of
particular functional groups. They must be performed in the
order specified because the functional groups tested for early
in the sequence may, if present, give spurious results in later
tests. For example, many amines give red precipitates with
dinitrophenylhydrazine reagent (test reagent for aldehydes and
ketones); they simply neutralize the acid in the reagent and so
cause the free base, 2,4-dinitrophenylhydrazine, to separate
from solution. Thus, amines must be shown to be absent before
aldehydes or ketones may be tested for. Similarly, many acids
interfere with tests for esters, aldehydes, ketones, alcohols,
etc.; esters interfere with aldehyde-ketone tests, and so on.
In elementary work, only one functional group normally will be
present, so that a characterization could stop with the first
positive indication. However, a little time spent completing
the full sequence will show some of the ways in which a functional group interferes with later tests. This information
and experience cannot be found in textbooks and is invaluable

in more advanced work with mixtures and with polyfunctional compounds.

Confirmatory tests are designed to confirm or negate the presence of a functional group suggested by the diagnostic test sequence. Although some confirmatory tests are quite specific, most will give positive results not only with the group sought but with several others also. Do not, in any circumstances, use a confirmatory test as a short cut to the diagnosis of a functional group. A negative confirmatory test usually is unambiguous; it indicates that the group suggested by the diagnostic sequence is, in fact, absent. A positive confirmatory test, on the other hand, only confirms that the group suspected may be present. The probability becomes virtual certainty if:

1) Several different confirmatory tests are consistent with the presence of the function indicated by the diagnostic sequence.

2) Diagnostic tests for all functions which might interfere with the confirmatory tests are negative.

3) Confirmatory tests for all possible interfering functions are negative. Obviously, one could spend considerable time on qualitative tests to convert the probability ever closer to certainty, but a point of diminishing returns soon is reached. You must learn to judge when you have diagnosed

and confirmed a functional group with sufficient certainty to justify the time and effort required to attempt the preparation of derivatives and final complete identification of the un-known substance.

The Reliability of Tests

Functional group characterization is based on the premise that a given group will exhibit its typical behavior irrespec-tive of the organic backbone to which it is attached. This premise simply is not valid. Although a large proportion of simple organic compounds do behave typically and predictably, it is safe to say that an absolute functional group test is impossible; that is, whatever the test or the group, one could devise substances containing the group which would test nega-tive and substances not containing the group which would give apparently positive results. The substances given as unknowns in introductory work are very carefully chosen to behave "normally" but even some of these can show anomalous behavior, usually for very straightforward reasons. Thus the simple amine, p-toluidine, does not dissolve readily in cold dilute sulfuric acid, not because it does not react but because its sulfate is rather insoluble in water. The base dissolves readily in dilute (2-5%) hydrochloric but not in concentrated acid because a high concentration of chloride ion lowers the solubility of the salt (common-ion effect). Similarly, a fair

number of aromatic acids will not dissolve in dilute sodium
hydroxide because their sodium salts are only slightly soluble,
and many more do not dissolve in concentrated alkali. Resolution of these particular anomalies is quite simple; use dilute
acid or base and test solubility in several different acids
(HCl, H_2SO_4, HNO_3, CH_3COOH, H_3PO_4) and bases (NaOH, KOH,
$Ca(OH)_2$, $Ba(OH)_2$, ammonia). Other apparent anomalies similarly
are quite readily and logically resolved. As a general rule,
using more of a reagent, or a more concentrated reagent,
unnecessarily, will not improve a test and usually will make
it worse! The ability to recognize and deal with anomalous
situations is a large part of the art of practical organic
chemistry.

Purity of Starting Materials

Substances issued as unknowns are ordinary commercial
samples. Most of these are reasonably pure, but many contain
significant amounts of impurities. This can present a trap
for the unwary or careless student. For example, commercial
benzaldehyde almost always contains a little benzoic acid and
therefore liberates carbon dioxide gas from bicarbonate solution. It does not, however, dissolve completely in cold
dilute alkali, although it will do so in hot concentrated
caustic alkali since it is converted thereby into benzoic acid
and benzyl alcohol(Cannizzaro reaction). Two excellent object-

lessons reside in this example; the student who takes a short
cut and trusts a confirmatory test (CO_2 from bicarbonate) but
ignores the diagnostic test (complete dissolution in dilute
alkali) will be faulted, as will also the student who uses unnecessarily powerful reagents or reaction conditions (hot concentrated alkali).

PROCEDURES

GENERAL PHYSICAL PROPERTIES

1. Obvious Physical Properties

Observe carefully and record as accurately as possible the
color, odor, physical state, general appearance, and any other
apparent characteristics of the unknown substance.

This simple examination, carefully considered, serves to
eliminate from consideration a vast number of possibilities.
For example, if the substance is liquid, all solids immediately
are discounted, and vice versa. Within the scope of this work
most alkanes, alkenes, ethers, aldehydes, ketones, alkyl
halides, aliphatic acids, and aliphatic amines are liquids.
Benzenoid acids are solids. Other compounds are liquids or
low-melting solids.

All compounds in these groups, when pure, are colorless.
Some, especially amines, phenols and iodine compounds often are
colored (commonly brown) by impurities. Distillation usually
removes the color.

131

4. FUNCTIONAL GROUP IDENTIFICATION

Most volatile compounds have rather characteristic odors.
Hydrocarbons and halogenated hydrocarbons are reminiscent of
gasoline, kerosene or naphtha. Amines smell of fish and
ammonia. Phenols have a characteristic "carbolic" or disin-
fectant odor while aliphatic acids and aldehydes are strong-,
sharp-, and often rancid-smelling. Esters have very characteristic pleasant fruity flavors (many are used as artificial
flavorings and perfumes) while ethers and ketones are less
fruity but still quite pleasant. Familiarize yourself with the
odors of all the common organic reagents and solvents in the
laboratory.

2. Approximate Melting or Boiling Point

Determine roughly the melting point and/or the boiling
point of the sample. Note that many low-melting solids
(phenols, amines, arylhalides, e.g.) boil at conveniently
attainable temperatures (below 250°) so that both properties
may be determined. Do not spend time and energy determining
these properties accurately at this stage, since the sample
issued is probably impure and the physical properties of impure substances are of little value for precise identification.
The approximate values, however, serve to sharpen the conclu-
sions suggested by the obvious physical properties, especially
in eliminating possible classes and possibilities within
classes. Thus, no benzenoid substance boils below 80°, no

alcohol below 65°, no acid below 100°, and no aryl halide below 130°. Most phenols and amines are liquids or low-melting solids and none boils below 180°. All aromatic acids are solids and virtually all of them melt above 100°. The common alkanes, alkenes, and aliphatic aldehydes, ketones, esters, ethers, amines and halogen compounds are liquids of low to moderate boiling point. Aromatic hydrocarbons, aliphatic alcohols and acids have moderate to high boiling points. Aromatic aldehydes, ketones, esters, ethers, alcohols, amines, phenols and halogen compounds are high boiling.

Study the tables in chapter 6 and familiarize yourself with the physical properties of the common members of the homologous series. Use your findings to confirm and supplement the above generalizations.

3. Solubility in Water

Add two drops of water to a small portion (one drop of a liquid; 20-30 mg of a solid) of the unknown substance. Agitate the mixture thoroughly and allow ample time (2-5 minutes) for dissolution to occur. Liquids, if soluble, usually dissolve rapidly, but many soluble solids dissolve rather slowly. Substances which dissolve completely under these conditions are regarded as being very soluble in, or miscible with water. Visualization of two liquid phases in a few drops of mixture often is difficult. If in doubt, withdraw the mixture into the

133

4. FUNCTIONAL GROUP IDENTIFICATION

nozzle of a long-nosed Pasteur pipet, when the interface between two liquids, if present, will be readily observable.

If dissolution does not occur in the cold, warm the mixture
in a hot water bath, again allowing ample time for dissolution.
If the substance dissolves, cool the solution thoroughly and
observe whether it separates again, either directly or, if a
solid, upon seeding with a minute trace of the original solid.
Low-melting solids will melt, of course, upon heating with
water; indeed, some solids, especially some phenols, liquefy
even in cold water but do not dissolve. Be very careful not
to report such behavior as dissolution unless one homogeneous
liquid phase actually results.

If the sample does not dissolve in two drops of water, add,
successively, a further three, five and ten drops, agitating
and warming the mixture and allowing ample time for dissolution
after each addition. Substances which dissolve under such conditions are regarded as being moderately to slightly soluble,
while those which do not dissolve in a total of twenty volumes
of water are considered insoluble.

Most organic compounds are but slightly soluble or insoluble in water. Water-solubility, then, is uncommon but, if observed, immediately restricts further consideration to a rather
small range of compounds, those having a significant proportion
of polar (OH, NH_2, C:O) groups attached to the non-polar

(hydrocarbon) backbone. Thus hydrocarbons, halogenohydrocarbons,
and most ethers, esters and phenols are but slightly soluble or
are insoluble. Monofunctional alcohols, acids, amines, aldehydes and ketones containing up to 3-5 carbon atoms are quite
soluble, often miscible in all proportions, and those containing 5-8 carbon atoms are moderately or slightly soluble. Above
this the hydrocarbon character predominates and the compounds
are insoluble unless two or more polar groups are present.
However, the main value of careful determination of water solubility is for comparison and contrast with solubility in dilute
mineral acids and alkalis (diagnostic tests 1 and 2 below).
The student who diagnoses that a substance is a base because a
portion (20 mg) did not dissolve in water (2 drops) but did
dissolve in dilute acid (20 drops) is clearly in trouble!

Acidity or basicity. Test the solution (or suspension)
of the sample with both blue and red litmus paper. If it
appears to be acidic add one drop of phenolphthalein indicator
solution followed, dropwise with shaking, by very dilute
(0.5%) sodium hydroxide solution. If several drops of the base
are required to turn the indicator then the substance is an
acid. If only one or two drops are needed the acidity is
probably due to trace impurities. Similarly, if the substance
is basic to litmus, determine its ability to neutralize very
dilute (0.5%) hydrochloric acid using congo red or methyl red

as indicator. This test is particularly reliable with soluble or moderately soluble substances for which it replaces diagnostic tests 1 and 2.

4. Detection of Elements Other than C, H and O

Nitrogen, chlorine, bromine and iodine in an organic substance are converted into the inorganic anions cyanide, chloride, bromide and iodide when the substance is heated with molten sodium. These are detected by the normal procedures of qualitative inorganic analysis.

Foreknowledge of the presence of these elements is not essential to a successful identification but is most desirable. It can be most valuable, for example, in confirming or negating a dubious test for a functional group containing such an element.

Procedure: Wear safety glasses and hold wire gauze as a screen between your face and the ignition tube. Place a small portion (10-30 mg of a solid, 1-2 drops of a liquid) of sample in the bottom of a small (10 x 70 mm) soft-glass test tube. A volatile liquid, or a substance suspected to contain nitrogen and little carbon, may be mixed with a little finely powdered pure sucrose or two drops of pure glycerol. Hold the tube at about 30° to the horizontal, mouth up, and carefully place two small pellets of clean sodium (about 1/8″ cube) about half-way down the tube. Hold the neck of the tube with tongs and apply

a **roaring** burner flame to the side of the tube so as to melt
the sodium without heating the sample appreciably, especially
if this is volatile. As soon as the sodium melts, turn the
tube to a vertical position so that the molten sodium falls
rapidly into the sample. Immediately, heat the bottom of the
tube strongly. Ignore fires and minor explosions, and heat the
lower half of the tube to red-heat for a minute or so (it will
crumple considerably). Allow the tube to cool, then hold it by
its lower end (with tongs!) and roast the upper half. This
will burn out any products of partial combustion which might
otherwise give spurious positive tests simulating halogen.
Finally, again holding the top of the tube, roast the lower
half again and plunge it, red hot, into 2-3 ml of distilled
water in a 10-25 ml beaker. The tube will disintegrate and its
contents (including any unconsumed sodium!) will dissolve in
the water. Discard the upper part of the tube, boil the aqueous suspension for a minute or two adding more water if necessary but avoiding undue dilution, allow to settle, decant or
pipet the solution into a centrifuge tube, and spin out insoluble material. If the supernatant solution is not clear, filter
it through a small (3 cm) filter paper in a small funnel. The
alkaline supernate or filtrate contains some of the nitrogen or
halogens of the original substance as cyanide or halides.

4. FUNCTIONAL GROUP IDENTIFICATION

Detection of Nitrogen: To about 0.5 ml of clear supernate add two or three small crystals of ferrous sulfate. Boil the solution for at least one minute (caution, it will froth), cool and acidify it with dilute hydrochloric acid. A dark blue precipitate (Prussian blue) indicates that nitrogen was present in the original substance.

$$Fe^{2+} + 6\ CN^- \xrightarrow{\text{boil}} [Fe(CN)_6]^{4-}$$

$$4Fe^{3+} + 3Fe(CN)_6^{4-} \longrightarrow Fe_4^{3+}[Fe(CN)_6]_3^{4-}$$
$$\text{Prussian blue}$$

Ample ferric ion invariably results during boiling; addition of ferric chloride is quite unnecessary. Occasionally, the blue pigment remains in colloidal suspension. Boiling, followed by centrifuging or long standing will coagulate and deposit the precipitate.

Detection of Halogen: Acidify 0.5 ml of clear supernate with dilute nitric acid. If the nitrogen test was positive, boil the acidified solution for several minutes to volatilize hydrogen cyanide. Do not, however, allow the solution to become too concentrated or halides may be oxidized by the nitric acid and lost. Cool the solution and add two drops of 1% aqueous silver nitrate. A white precipitate, insoluble in dilute nitric acid but soluble in dilute ammonia indicates chlorine. Silver bromide is cream or off-white and difficultly soluble in ammonia. A light yellow precipitate, insoluble in ammonia indicates iodine.

Confirmation of Iodine and Bromine. (a) Many organic iodine and bromine compounds give elemental iodine or bromine (characteristic violet or brown vapors respectively) when a small portion (10 mg) is heated with concentrated sulfuric acid (one drop) and a little manganese dioxide. (b) Add 0.5 ml of carbon tetrachloride to an acidified (dil. HNO_3) but unboiled portion (0.5 ml) of fusion supernate. Then add dilute sodium hypochlorite solution dropwise, while shaking the mixture thoroughly. Ensure that the mixture remains acidic. Iodine or bromine liberated will dissolve in the lower (CCl_4) layer and impart to it a characteristic violet or brown color. Excess hypochlorite destroys either color.

DIAGNOSTIC SEQUENCE

1. Solubility in Dilute Mineral Acid (2-5% HCl, H_3PO_4, HNO_3, H_2SO_4): Diagnosis of Organic Base

This test is applied only to substances which are insoluble or but slightly soluble in water. Add hydrochloric acid (2-5%) dropwise, with shaking, to a few mg or one drop of sample. Give ample time after each drop for dissolution to occur, es- pecially with solid substances. If the substance dissolves ob- serve whether it is reprecipitated when the solution is neutralized by careful, dropwise addition of dilute (5%) alkali. If reprecipitation does not occur it is probable that the substance simply dissolved in the water; compare carefully with the solubility in water alone. If the substance does not dissolve in hydrochloric acid, repeat the test with one or more of the other acids listed.

Substances which are insoluble (or slightly soluble) in water but dissolve in dilute acids are obviously bases. Most organic bases contain nitrogen and within the scope of this work the only bases encountered will be amines. The test should be performed even on substances in which nitrogen was not found; if it is positive the nitrogen test should be repeated most carefully.

The salts resulting from reaction of most simple amines with common acids are quite soluble, but some are not. Hence, if the substance does not dissolve in dilute hydrochloric, the other acids should be tried also. It is very improbable that the salts of any given amine with all the four acids listed will be insoluble. Dilute nitric acid often contains some nitrous acid. This should be destroyed by adding a little urea to the acid before use (why?). Most amines are much less soluble in concentrated or moderately concentrated than in dilute acids (common-ion effect). Some amines, on the other hand, are such weak bases that they will only dissolve in moderately

concentrated acid (15-20% HCl, 25% H_2SO_4 or H_3PO_4). Many substances other than amines will dissolve in acids at such concentrations (alcohols, some ethers, e.g.) so that such dissolution should only be tried when dilute acid has failed, and the
results must be interpreted carefully.

2. Solubility in Dilute Alkali (2-5% NaOH, KOH, $Ca(OH)_2$,
$Ba(OH)_2$, ammonia): Diagnosis of Acid Substances

This test is applied only to substances which are insoluble
or but slightly soluble in water. Add dilute (2-5%) sodium
hydroxide solution dropwise, with shaking, to a few mg or one
drop of sample. Give ample time after each drop for dissolution to occur. If the substance dissolves, observe whether it
is reprecipitated upon neutralization with dilute acid. If the
substance does not dissolve in sodium hydroxide, repeat the
test with one or more of the other bases.

Substances which dissolve in dilute alkali but not in water
are classified as acidic substances. Two kinds only, carboxylic
acids and phenols, are commonly encountered. These are confirmed and differentiated by confirmatory tests.

Common-ion effects contraindicate the use of concentrated
base (except, perhaps, ammonia). Few acids or phenols are so
weakly acidic that concentrated alkali is ever necessary.

4. FUNCTIONAL GROUP IDENTIFICATION

3. <u>Reaction</u> <u>with</u> <u>Hot</u> <u>Alcoholic</u> <u>Alkali</u>: <u>Diagnosis</u> <u>of</u> <u>Ester</u>

Place a portion (10-20 mg or 1 drop) of the unknown substance in one of two test tubes. Add to each tube two drops
of ethanol, one drop of phenolphthalein solution and one drop
<u>only</u> of <u>dilute</u> (2%) sodium hydroxide. The contents of both
tubes should be bright pink (unless the substance is an acid or
a phenol); if either is not, add further sodium hydroxide and
ethanol (equal volumes) dropwise until they are. Stopper both
tubes loosely and heat them at 100° (boiling water bath) for
up to 30 minutes, and observe any color changes. If the color
of the test sample is discharged but that of the blank is not,
a functional group (ester) is present which hydrolyzes slowly
and neutralizes the alkali. If the blank becomes colorless,
add a further drop of phenolphthalein to both tubes and continue the heating.

A variety of substances other than esters (amides,
nitriles, acid anhydrides and acid halides,for example) also
give positive results. These are outside the present scope,
but this test is diagnostic for all of them and is so used in
later work. They are differentiated by appropriate confirmatory
tests. Certain aldehydes, especially benzenoid aldehydes,
react similarly since they disproportionate to give the corresponding acids and alcohols(Cannizzaro reaction). These,
usually, are readily differentiated by their behavior with

142

dinitrophenylhydrazine (test 4, below) and by confirmatory tests.[†]

The test should be performed even on a substance known to contain an acidic function since bifunctional compounds containing both acid and ester groups (acetylsalicylic acid, aspirin and methyl salicylate, oil of wintergreen, e.g.) will be encountered early in moderately advanced work.

If the test is positive, the hydrolysis should be taken to completion. To the tube containing the partly-hydrolyzed sample add five drops of 20% sodium hydroxide and five drops of ethanol. Heat at 100° for 30-60 minutes. The fruity odor (if any) of the ester should disappear completely. Cool the solution and dilute it with water (1 ml). No precipitate or oil should separate unless the ester is incompletely hydrolyzed or gives rise to a water-insoluble alcohol. Acidification (20% HCl) of the diluted hydrolysate may generate the characteristic odor of a volatile acid (acetic, butyric) or give a precipitate of an insoluble one.

[†]Thus certain aldehydes test positive for ester and some esters give precipitates with 2,4-dinitrophenylhydrazine. It is a moot point, therefore, whether the ester test or the aldehyde-ketone test should be performed first. Rarely is there difficulty in differentiation, however, since most esters react very slowly (if at all) with dinitrophenylhydrazine, and the confirmatory hydroxamate and Tollens' tests reliably negate ester and aldehyde, respectively.

4. FUNCTIONAL GROUP IDENTIFICATION

These observations are of great value in choosing reactions for final characterization of the ester. Further work need not be delayed during the hour required for the hydrolysis.

4. Reaction with 2,4-Dinitrophenylhydrazine: Diagnosis of Aldehyde or Ketone

To one small drop (no more) of the sample, add 2-3 ml of 2,4-dinitrophenylhydrazine reagent. Shake the mixture thoroughly. If no precipitate results within 2-3 minutes, divide the solution and heat one portion in a hot bath (50-80°; do not boil) for five minutes, cool, shake, and rub the inside of the tube at the liquid surface with a glass rod. Finally, if no precipitate is observed, set both tubes aside for 30 minutes.

Most aldehydes and many simple ketones give orange or red precipitates within ten minutes at room temperature. Less-reactive aldehydes and most ketones react upon heating; the derivative often remains dissolved but crystallizes upon cooling and rubbing. If the reagent is boiled, small amounts of a red precipitate often result due to evaporation of the solvent (methanol) or redox disproportionation of the nitrohydrazine. Do not use too much of the unknown or a bright orange liquid, floating on the reagent solution, will result. This is indistinguishable from the result observed with almost any water-insoluble substance.

Saturated carbonyl compounds typically give yellow or orange derivatives. Unsaturated ones give orange, orange-red or red products. The product should be isolated for crystallization to purity as one crystalline derivative of the unknown substance.

The test is quite specific if done properly. A few amines, acids and esters (acid anhydrides, acid halides, amides) may interfere, especially if proper conditions are not adhered to (too much unknown is the commonest reason) but such interference is rare and readily rationalised.

5. Metallic Sodium Test: Diagnosis of Lower Alcohol

This test is applied only to liquids. The test tube and the pipet must be absolutely dry. To a minute piece of clean, freshly cut sodium (the size of a pin's head) add one drop of the liquid substance. If the substance is a lower alcohol, hydrogen gas will be evolved steadily until the sodium is completely dissolved.

A brief evanescent effervescence should be discounted since most liquid substances contain traces of water which react, of course, until consumed. Amines, carboxylic acids, and phenols react as do alcohols so that the test should not be applied to substances containing these functions. Many esters, aldehydes and ketones also react, some much like alcohols

4. FUNCTIONAL GROUP IDENTIFICATION

(what reactions are involved?) so that the test can only be used negatively if these functions are present.

6. <u>Solubility</u> <u>in</u> <u>Cold</u> <u>Concentrated</u> <u>Sulfuric</u> <u>Acid</u>: <u>Differen-tiation</u> <u>of</u> <u>Alcohols</u> <u>and</u> <u>Alkenes</u> <u>from</u> <u>Ethers</u> <u>and</u> <u>from</u> <u>Hydrocarbons</u> <u>and</u> <u>Halogeno-hydrocarbons</u>

Chill thoroughly in an ice-water bath a test tube contain-ing two drops (10 mg of a solid) of the unknown. Add ten drops of concentrated sulfuric acid, allowing each drop to drain slowly down the chilled inner wall of the tube and dissipating any heat evolved by shaking the tube in the bath before adding the next drop. Observe whether the substance dissolves. With solids dissolution may take some minutes; with liquids it may be difficult to distinguish, in which case the mixture should be completely withdrawn into a <u>dry</u> pipet, when the interface between two layers, if present, will be clearly visible. If the substance dissolves, dilute the cooled solution with ice-water (0.5 ml) and observe whether it is reprecipitated. Do not allow the mixture to get hot or side reactions and dis-coloration will vitiate the test.

Simple alcohols, alkenes and ethers dissolve in the acid. Alkanes, simple benzenoid hydrocarbons, alkyl and aryl halides (unless they contain alcohol, alkene or ether functions also) do not. Ethers are reprecipitated upon dilution. Alcohols and alkenes are not, since they are converted into the soluble

146

alkyl hydrogen sulfates. Alcohols are differentiated from alkenes by means of confirmatory tests (acetyl chloride, bromine in carbon tetrachloride, Baeyer's reagent), halides from hydrocarbons by elemental composition and aliphatic from aromatic compounds by the non-smoky or smoky flame upon combustion and the general physical properties.

Many moderately complex alcohols, alkenes, ethers and benzenoid hydrocarbons do not exhibit precisely the simple behavior described above. For example, benzyl alcohol, styrene, and methyl benzyl ether polymerize, while reactive benzenoid hydrocarbons such as mesitylene (1,3,5-trimethyl benzene) and anthracene are converted slowly into sulfonic acids which dissolve and do not reprecipitate upon dilution. However, careful application of the test, coupled with some common sense consideration of the actual observations, the general physical properties and a few appropriate confirmatory tests usually leave no doubt as to the nature of the compound. Indeed, anomalous behavior, properly interpreted, often gives much more information than does simple behavior Thus, using the examples cited above, benzyl alcohol, styrene, and methyl benzyl ether are converted by sulfuric acid into insoluble solid or viscous-liquid polymers. If such behavior is observed, the substance obviously is not a saturated or benzenoid hydrocarbon or halide (even a readily sulfonated one). Confirmatory tests readily distinguish between the three remaining possibilities.

Extreme care must be taken, however, to ensure that all functional groups up to and including aldehydes and ketones have been properly diagnosed before testing with sulfuric acid since all of these functions will react, some in most peculiar and dramatic ways.

7. Combustion: Differentiation of Benzenoid from Non-benzenoid Compounds

Set fire to a portion (2 drops or 20 mg) of the unknown substance on a spatula or crucible lid. Observe whether or not it burns with a smoky flame. Compare with known substances (hexane, ethanol, benzene, napthalene, etc.). Generally, substances containing much carbon and relatively little hydrogen, most typically the benzenoid compounds, produce much smoke when burned; substances containing more hydrogen do not. There are many exceptions, of course; polyhydric phenols burn without smoke while highly unsaturated aliphatic compounds (lower al-kynes, dienes) and higher saturated substances (stearic acid) burn smokily, but the test is very simple, useful and usually quite unambiguous when coupled with other observations.

CONFIRMATORY TESTS

1. Amines

Reaction with Nitrous Acid. Dissolve a portion (20 mg or one drop) of the suspected amine in 2-5% hydrochloric or phosphoric acid (2-4 ml). Cool the solution in an ice bath. The amine salt may crystallize from solution. Add to the cooled mixture 2-3 ml of 1% sodium nitrite solution or 20-30 mg of solid $NaNO_2$.

(a) Separation of a pale yellow oil or low melting solid (N-nitrosoamine) which does not dissolve upon warming to room temperature indicates a secondary amine.

$$\underset{R}{\overset{R}{>}}\overset{+}{N}\underset{H}{\overset{H}{<}} + H-O-N=O \longrightarrow \underset{R}{\overset{R}{>}}N-N=O + H_3O^+$$

N-Nitrosoamine

Note, however, that the lower aliphatic primary amines give water-soluble nitrosoamines.

(b) Vigorous evolution of gas (nitrogen) indicates a primary aliphatic amine.

$$R-NH_3^+ + H-O-N=O \longrightarrow R-OH + N_2 + H_3O^+$$

Comparison with a blank is essential since the reagents alone evolve some gas. A more reliable test is the evolution of nitrogen upon mixing solutions of the amine and sodium nitrite in 80% sulfuric acid.

4. FUNCTIONAL GROUP IDENTIFICATION

(c)   Add the cold reaction solution to a solution of
2-naphthol (10-20 mg) in 10% sodium hydroxide (4 ml).   Formation of an orange or red azo dye indicates a primary aromatic
amine.

$$Ar-NH_3^+ + H-O-N=O \longrightarrow Ar-N_2^+ + 2H_2O$$

$$Ar-N_2^+ + \quad \longrightarrow \quad Ar-N=N-$$

Arylazo-2-naphthol

(d)   No apparent reaction indicates that the substance
either was a tertiary amine or not an amine at all.

Hinsberg's Test.   Suspend the suspected amine (20 mg or one
drop) in 20% sodium hydroxide (1-2 ml) and add two drops of
benzenesulfonyl chloride (or 50 mg of p-toluenesulfonyl chloride).   Stopper the tube and shake the mixture vigorously until
the smell of the sulfonyl chloride has disappeared (5-10
minutes).

(a)   Primary amines form sulfonamides which remain dissolved in the strongly alkaline solution.

$$R-NH_2 + Cl-SO_2- \xrightarrow{OH^-} R-NH-SO_2- \xrightarrow{OH^-} R-\overset{-}{N}-SO_2-$$

Acidification precipitates the white solid sulfonamide.
(Caution: From such concentrated solutions, sodium chloride
may separate upon acidification. Ascertain that any white
solid precipitate is, indeed, insoluble in water and dilute
acid and soluble in 10-20% alkali).

(b) Secondary amines form sulfonamides which do not remain
in solution but precipitate directly from the alkaline reaction
mixture as white solids, insoluble in water, dilute alkali and
dilute acid.

(Caution: If the sodium hydroxide is too dilute, or if too
much sulfonyl halide is used, a primary sulfonamide may precipitate; if it is too concentrated, the sodium salt of a
primary sulfonamide, or even sodium chloride, may precipitate.
Ascertain that any precipitate from the alkaline reaction mixture is, indeed insoluble in water, dilute alkali and concentrated alkali.)

(c) Tertiary amines do not react, remain undissolved, and
dissolve upon acidification.

If a solid product is obtained from a primary or secondary
amine, isolate this and keep it for crystallization to purity
as a derivative.

4. FUNCTIONAL GROUP IDENTIFICATION

2. Carboxylic Acids and Phenols

Liberation of Carbon Dioxide from Sodium Bicarbonate.
Dissolve 50 mg of solid sodium bicarbonate in one ml of water.
Allow any bubbles to disappear. If the suspected acid is soluble or moderately soluble in water, add the bicarbonate solution gently to one drop (20 mg) and agitate gently. If the
sample is insoluble or slightly soluble in water, dissolve it
in one ml of ethanol or acetone and add the bicarbonate solution to this solution. Precipitation, which will occur, should
be ignored. Look for vigorous effervescence or steady evolution of gas.

Carboxylic acids and a few unusually strongly acidic phenols (typically nitrophenols, which invariably are colored or
give colored solutions in dilute alkali) liberate CO_2. Most
phenols, however, do not.

Solubility in Dilute Sodium Carbonate Solution. The test
is performed as is that for solubility in water, dilute acid or
alkali. Most carboxylic acids are soluble, as they are in
sodium hydroxide. Most phenols (except strongly acidic ones)
are not.

Color with Aqueous Ferric Chloride. Add one ml of dilute
ferric chloride solution to a very small amount of a liquid
unknown, or to a solution of 5 mg of a solid unknown in two
drops of ethanol. Shake the mixture. Most phenols give

vividly colored solutions (blue, green, violet, etc.). Acids
(except phenolic acids)do not, although some give yellowish-
buff solutions or precipitates. *—olive green - minute amt. ppt.*

Color with Ferric Chloride-Pyridine in Chloroform. Dissolve a very small drop (5 mg) of the unknown in 1 ml of a 0.1%
solution of pyridine in chloroform. Add one ml of ferric chloride in chloroform and shake the mixture. Virtually all phenols, and especially those which do not give colors with aqueous ferric chloride, give colored solutions. Carboxylic acids,
again, do not.

3. Esters

Hydroxamate-Ferric Color Test. Dissolve half a pellet
(50 mg) of potassium hydroxide in one drop of water, add one
ml of ethanol, 20-30 mg of hydroxylamine hydrochloride and one
drop of the suspected ester. Heat the solution for 3-5 minutes
in a boiling water bath, adding ethanol if necessary to make
good any loss by evaporation. Cool, acidify the solution with
dilute hydrochloric acid, and add two drops of aqueous ferric
chloride. The development of a brilliant rose-red color confirms ester.

$$R\text{-}\overset{\overset{\displaystyle O}{\|}}{C}\text{-}O\text{-}R' + NH_2OH \longrightarrow R\text{-}\overset{\overset{\displaystyle O}{\|}}{C}\text{-}NH\text{-}OH \xrightarrow{Fe^{3+}} \text{red complex}$$

153

4. FUNCTIONAL GROUP IDENTIFICATION

4. Aldehydes and Ketones

The classification test with dinitrophenylhydrazine, properly performed, is so reliable that confirmation is rarely necessary. The following tests are useful occasionally.

Tollens' Test. Add dilute ammonia dropwise to 0.5 ml of 5% aqueous silver nitrate in a clean test tube until the dark brown precipitate of silver oxide initially formed just redissolves. Do not use excess ammonia. Add to the solution the smallest possible amount of a liquid (especially if this is a water-insoluble liquid) or 5 mg of a solid unknown, shake thoroughly and warm the solution to 50-60° (no hotter) in a water bath for several minutes. The formation on the inside wall of the tube of a silver mirror (good enough to shave by; not a dirty brownish-silver smudge) indicates aldehyde. Excess ammonia, too much unknown, overheating, or a dirty tube, will all give poor results.

Upon completion of the test wash away the reaction mixture with plenty of water since it may deposit explosive products upon standing or drying out. The mirror may be removed with hot nitric acid.

Iodoform Test. Crush a large crystal of iodine (100 mg) and dissolve it in a little dilute aqueous sodium hydroxide. To the colorless solution, add two drops of the unknown substance. Heat the mixture in a hot water bath for several minutes. Separation of pale yellow crystals of iodoform and development of the very characteristic odor of this substance indicates a methyl ketone (or acetaldehyde, or an alcohol capable of oxidation to either of these). For experience, try the reaction with an authentic methyl ketone (acetone or aceto-phenone).

$$R-\underset{\underset{O}{\|}}{C}-CH_3 \xrightarrow{3IO^-} R-\underset{\underset{O}{\|}}{C}-CI_3 \xrightarrow{H_2O} R-\underset{\underset{O}{\|}}{C}-OH + H-CI_3$$

5. Alcohols

Reaction with Acetyl Chloride. This test is particularly useful for confirming lower aliphatic alcohols which, being soluble in water, are not diagnosed by reaction with cold concentrated sulfuric acid.

$$R-OH + CH_3-CO-Cl \longrightarrow R-O-CO-CH_3 + HCl$$
$$\text{Ester}$$

Place 3-4 drops of acetyl chloride in a dry test tube. Allow any fumes resulting from reaction of the acid chloride with atmospheric moisture to dissipate, and then add, one drop at a time, three drops of the suspected alcohol. A positive indication is given by:

(a) A vigorous reaction; the mixture boils spontaneously.

(b) Heat of reaction; the mixture becomes warm or hot (hold the bottom of the tube on the back of your hand).

4. FUNCTIONAL GROUP IDENTIFICATION

(c) Hydrogen chloride gas is evolved; detect this by
holding a glass rod wetted with concentrated ammonia near the
mouth of the tube, when dense white fumes of ammonium chloride
will result. (Caution: Acetyl chloride is volatile and may
also give white fumes; compare with a blank). Cool the mixture
in ice, add several drops of water and shake the mixture to
destroy any unconsumed acetyl chloride. Make the mixture alkaline with dilute sodium hydroxide solution and observe whether
the mixture now has the pleasant fruity smell of a volatile
ester.

6. Alkenes

Decolorization of Bromine. To 2 drops of the suspected
alkene in a dry test tube, add dropwise a 1% solution of bromine in carbon tetrachloride. An alkene will decolorize
rapidly at least ten drops of the bromine solution.

$$\text{>C = C<} + Br_2 \longrightarrow \underset{Br \quad Br}{\text{>C - C<}}$$

Most commercial samples of alkanes contain traces of alkenes, sometimes sufficient to decolorize one or two drops of
bromine solution but no more. Substances such as phenols,
phenol ethers, some alcohols, aldehydes, ketones, and a few
highly reactive aromatic hydrocarbons will decolorize bromine,
usually rather slowly. In all these cases hydrogen bromide is
produced. This may be detected by holding a glass rod, wetted

with concentrated ammonia, close to the mouth of tube, when dense white fumes of ammonium bromide will be observed. Alkenes do not give hydrogen bromide when they react with bromine.

Baeyer's Test. Add 0.1% aqueous potassium permanganate solution dropwise to two drops of the suspected alkene, shaking the mixture vigorously. An alkene will discharge the purple color of at least 1 ml of the permanganate solution, and a brown precipitate of manganese dioxide will result.

Many commercial samples of alkanes contain traces of alkenes, and so will decolorize a few drops of the permanganate. Aldehydes, amines, some phenols and other reducing substances also reduce permanganate giving MnO_2. Many lower alcohols react (usually rather slowly) especially if they contain aldehydes as impurities.

7. Ethers

Ferrox Color Test. Add 2-3 drops of the substance to a small piece (1/4" square) of Ferrox[+] test paper in a dry test

---

[+]Ferrox test paper is prepared as follows: Dissolve 1 g of $FeCl_3.6H_2O$ and 1 g of KCNS separately in 10 ml portions of methanol. Mix the solutions and filter. Dip strips of filter paper in the solution 2 or 3 times, drying them in the air between dips. The resulting dry test paper, which should exhibit a greenish-metallic sheen, should be stored in the dark.

tube. If the liquid is an ether, it will become red due to ferric thiocyanate dissolved out of the paper. A solid ether may be dissolved in the minimum amount of dry benzene, and the test applied to the solution.

This test distinguishes compounds containing oxygen (including ethers) from those which do not (hydrocarbons and halogeno-hydrocarbons). Thus acids, alcohols, aldehydes, ketones and esters, as well as ethers, give positive results. So does water. If it is suspected that the sample may not be absolutely dry, it should be dried with anhydrous magnesium sulfate or sodium sulfate before testing.

8. Halogen Compounds

Mobility of Halogen. The test determines the ease with which the halogen atom separates from the organic compound as halide ion. This is governed by the structure to which the halogen is attached and so gives some indication thereof. It also indicates whether, and under what conditions, the halogen group can be used as a functional group for the preparation of derivatives.

(a) If the halogen-containing substance is high-melting or moderately high-melting (above 100°) and is soluble or fairly soluble in water it may be the hydrohalide salt of an amine and contain ionised halogen. Dissolve a few milligrams of the substance in 0.5 ml of dilute acetic or nitric acid. If it does not dissolve completely, warm the mixture for a few minutes, cool, and separate the supernatant liquid. To the cool clear solution, add several drops of 1% aqueous silver nitrate solution. Immediate precipitation of silver halide indicates ionised halogen. Ascertain that the precipitate is insoluble in water, dilute nitric acid, alcohol and acetone. Silver chloride (white) is soluble in dilute ammonia. Silver iodide (yellow) is not.

A few substances containing very reactive covalently bound halogen (acid chlorides and iodides, α-halogeno acids, α-halogeno ethers) will also give precipitates of silver halide under this conditions.

(b) If test (a) is negative or is unnecessary (substances giving positive results in test (a) are not encountered in elementary work) dissolve the substance (1 drop or 20 mg) in ethanol and add two drops of 1% alcoholic silver nitrate solution. Lower alkyl iodides and bromides, allyl and benzyl halides (including chlorides) and sulfonyl and acyl halides give precipitates of silver halide in the cold within a few minutes. Upon heating the mixture, higher alkyl iodides and bromides and many simple alkyl chlorides react completely or partially, as do activated aryl halides such as 2,4-dinitrochlorobenzene.

(c) If test (b) is negative, dissolve the substance (1 drop or 10 mg) in ethanol (0.5 ml), add 5 drops of 20% sodium

159

hydroxide solution and heat the mixture in a boiling water bath
for several minutes. Cool, acidify with dilute nitric acid,
boil (to eliminate CO_2), cool, and add aqueous silver nitrate.
Alkyl chlorides give precipitates of silver chloride.

(d) If no silver halide results under any of the above
conditions, the halogen (if present) is attached directly to a
benzene ring (aryl halide) or to an ethylenic carbon atom
(vinyl halide).

Approximate Reactivity Order	Test
1. Hydrohalide salts of amines	(a)
2. Acid chlorides	(a), (b) - cold
3. α-halogeno acids and ethers	(a), (b) - cold
4. Allyl halides	(b) - cold
5. Sulfonyl halides	(b) - cold
6. Benzyl halides	(b) - cold (slowly)
7. Lower alkyl iodides	(b) - cold (slowly)
8. Higher alkyl iodides	(b) - hot
9. Lower alkyl chlorides	(b) - hot, (c)
10. Activated aryl halides	(b) - hot, (c)
11. Higher alkyl chlorides	(c)
12. Aryl halides	No reaction
13. Vinyl halides	No reaction

Caution: Very many commercial samples of halogen compounds
contain traces of the free halogen (iodides) or the halogen

hydracid and so give <u>slight</u> precipitates of silver halide immediately. If this occurs, compare the amount of silver halide produced under the conditions of test (a) with that produced in test (b) or (c), and draw the appropriate conclusion.

5

SELECTION AND PREPARATION OF SOLID DERIVATIVES

"A deceased body warps the mind."

"We believe that reptiles came from amphibians by spontaneous generation and the study of rocks."

"The earth makes a resolution every twenty-four hours."

The physical and chemical properties of an exceedingly large number of organic substances have been determined and recorded. Lists of these physical properties, such as those in Chapter Six of this book and the more comprehensive lists in various compendia, make this information readily available.

The final unique identification of a substance involves determining the physical properties of the substance itself and of a number of other substances into which the substance has been converted, its derivatives. If all the properties determined agree with those of one compound and no other, one may affirm not only that the substance is identified within the scope of the literature consulted, but also, if a sufficient number of properties are determined, that the possibility of its being any other substance is vanishingly small.

Conversion into crystalline solid derivatives affords a most useful identification procedure. Substances must be pure

5. SELECTION AND PREPARATION OF SOLID DERIVATIVES

before their physical properties are determined - a physical property of an impure substance is about as useful as a smudged fingerprint - and crystallization from solution is by far the easiest and most reliable purification procedure for small quantities of material. And a melting point is reliably and readily determined.

A substantial investment of time and effort is involved in attempting to prepare and purify a derivative. Be very sure, therefore, that the functional group to be reacted really is present.

Select derivatives carefully. Consider all the information gleaned from functional group tests and the approximate physical properties of the unknown substance in conjunction with the tables so as to narrow the field of possibilities, and choose derivatives which will afford the maximum possible differentiation between those which remain. Choose derivatives which will be solid, of course, and wherever possible select those which melt above 60-80°; lower-melting substances often are troublesome to crystallize.

There is no "best" derivative for any class of compound, or even for any compound. The author has exercised personal preference in listing the methods of preparation. Generally, those detailed first for each class of compound are most reliably produced and purified, display the widest range of melting points, and are well chronicled.

5. SELECTION AND PREPARATION OF SOLID DERIVATIVES

Certain substances, particularly the aliphatic hydrocarbons and ethers, are not conveniently converted into crystalline derivatives. Identification of these depends on the
physical properties, particularly the refractive index, the
density and the boiling point of the unknown substance itself.
This must be carefully purified, usually by distillation, if
reliance is to be placed on the data obtained. The refractive
index is a convenient characteristic of many other noncorrosive
liquid substances such as alcohols, esters and aromatic hydrocarbons. They must be dried and purified, of course, and at
least one derivative should be prepared.

There is no specific body of data or number of derivatives
which will identify an unknown substance. The required statement is: "The likelihood that any substance other than that
named might exhibit all the properties observed is negligible."
Occasionally, one derivative (or precise numerical physical
property such as refractive index) together with an overwhelm-
ing body of very characteristic qualitative observations will
suffice, but two derivatives are usually necessary and three or
more are sometimes needed. Bear in mind that the tables of
compounds in this book, and even those in larger compendia, are
far from comprehensive. If in doubt, especially in advanced
work, make at least one more derivative than is needed to identify the substance uniquely within the tables considered.

165

5. SELECTION AND PREPARATION OF SOLID DERIVATIVES

METHODS OF PREPARATION OF SOLID DERIVATIVES

DERIVATIVES OF PRIMARY AND SECONDARY AMINES

1. p-Toluenesulfonyl and Benzenesulfonyl Derivatives
   (N-substituted Sulfonamides)

   (a) Schotten-Baumann Procedure.

$$R_2NH + Cl-SO_2-C_6H_5 + OH^- \longrightarrow R_2N-SO_2-C_6H_5 + H_2O + Cl^-$$

Shake vigorously for ten minutes (or until the smell of sulfonyl halide disappears) in a stoppered test tube a mixture of the amine (100 mg) with 10% sodium hydroxide (5 ml) and p-toluenesulfonyl chloride or benzenesulfonyl chloride (250 mg). Acidify with dilute hydrochloric acid, separate the solid product by centrifuging, decant or pipet off the supernatant liquid and drain the product. Wash it twice with water or very dilute hydrochloric acid. Crystallize from methanol or from ethanol, or from a mixture of either of these solvents with water.

The procedure is much as in the Hinsberg test. If this test has already been carried out successfully, sufficient product may be available for crystallization to purity.

   (b) Pyridine Procedure. This procedure is somewhat less convenient than the Schotten-Baumann method, but gives better results with weakly basic amines such as nitroamines. Heat the amine (100 mg) with p-toluenesulfonyl chloride or benzene-

sulfonyl chloride (250 mg) and pyridine (1 ml) at about 100°
for 30-60 minutes. Pour the mixture into 2-3 ml of cold water
and stir until the product solidifies. Isolate, wash and crystallize it as under (a) above.

2. Benzoyl and Acetyl Derivatives (N-substituted Benzamides
   and Acetamides)

$$R_2NH + Cl-\overset{\overset{O}{\|}}{C}-C_6H_5 + OH^- \longrightarrow R_2N-\overset{\overset{O}{\|}}{C}-C_6H_5 + H_2O + Cl^-$$

$$R_2NH + CH_3-\overset{\overset{O}{\|}}{C}-O-\overset{\overset{O}{\|}}{C}-CH_3 + OH^- \longrightarrow R_2N-\overset{\overset{O}{\|}}{C}-CH_3 + CH_3COO^- + H_2O$$

Shake vigorously for ten minutes (or until the smell of
benzoyl chloride or acetic anhydride disappears)in a stoppered
test tube a mixture of the amine (100 mg) with 10% sodium
hydroxide (5 ml) and acetic anhydride or benzoyl chloride (5
drops). Separate the solid product by centrifuging (do not
acidify), decant or pipet off the supernatant liquid and drain
the product. Wash it twice with water and crystallize it from
methanol, ethanol, or from a mixture of either of these solvents with water.

3. N-substituted Phthalimides   (Primary Amines only)

Heat under reflux for 20-30 minutes a mixture of the primary amine (100 mg),glacial acetic acid (1 ml) and phthalic anhydride (100 mg). The product crystallizes upon cooling. Crystallize it from ethanol or glacial acetic acid.

5. SELECTION AND PREPARATION OF SOLID DERIVATIVES

4. Formyl Derivatives (N-substituted Formamides)

Heat under reflux for ten minutes a mixture of the amine (100 mg) and 90% formic acid (1 ml). Dilute the hot solution with cold water (2 ml) and cool in ice. If the derivative does not separate, saturate the solution with salt. Wash the solid product with water and crystallize from water, alcohol, aqueous alcohol or (after thorough drying) from petroleum ether b.p. 60-80°.

5. Picrates (Salts or Molecular Complexes with Picric Acid) (Primary, secondary and tertiary amines)

Suspend the amine (100 mg) in cold saturated (1% aqueous) picric acid (2,4,6-trinitrophenol). If the amine is solid, heat the mixture to about the m.p. of the amine or to 100° (boiling-water bath), whichever is lower. Shake the mixture vigorously (and allow it to cool, if hot). If no product crystallizes, allow the mixture to stand for 10-30 minutes, shake again, and rub the inside of the vessel with a glass rod to induce crystallization. Recrystallize from water, ethanol, aqueous ethanol, aqueous acetic acid, or (after drying) from benzene.

Caution: Picric acid and its salts are high explosives. Many of its heavy-metal salts (which may form on contact with iron, lead or copper articles such as drainpipes) are detonators also. Wash away waste picric acid and its compounds with generous quantities of water.

DERIVATIVES OF CARBOXYLIC ACIDS

1. Determination of Equivalent Weight

The equivalent weight of an acid, accurately determined (±1% or better) is fully as valuable as a derivative. It is readily determined for virtually any acid. The acid must be pure, of course. If there is any doubt as to its purity, it should be recrystallized or distilled before the determination is made.

Weigh accurately (analytical balance) a clean dry 100 ml Erlenmeyer flask. If the acid is volatile (smell) the flask should be stoppered (plastic stopper) and weighed with its stopper. Place a sample of the acid (about 200 mg) in the flask, restopper if appropriate, and reweigh accurately. The difference in weight is the weight of the acid. Dissolve the acid by adding 10 ml of water or (if it is insoluble) ethanol, add two drops of phenolphthalein indicator solution, and titrate with standard sodium hydroxide (0.1 normal) to a _faint_ pink end-point.

$$\text{Equivalent weight} = \frac{\text{milligrams of acid}}{\text{Ml of alkali x normality}}$$

The equivalent weight is equal to the molecular weight for a monobasic acid and to a simple fraction (one-half, one-third, etc.) thereof for a polybasic (dibasic, tribasic, etc.) acid.

5. SELECTION AND PREPARATION OF SOLID DERIVATIVES

2. Anilides and p-Toluidides (Carbodiimide Procedure)

Add to a solution of the acid (100 mg) and aniline or p-toluidine (100 mg) in dry chloroform, carbon tetrachloride or ethyl acetate (2-3 ml) a solution of dicyclohexylcarbodiimide (100 mg) in a similar volume of the same solvent. Warm the solution briefly (do not boil) and allow it to stand for 15-30 minutes or until precipitation of dicylohexylurea is complete. Filter the suspension, wash the solid urea with three 2-ml portions of warm solvent, and combine and evaporate the filtrate and washings. Extract the resulting anilide or p-toluidide successively with aqueous sodium carbonate (5%), hydrochloric acid (5%) and water. Crystallize it from ethanol, methanol, water or aqueous alcohol.

3. p-Bromophenacyl, p-Phenylphenacyl and p-Nitrobenzyl Esters

Dissolve or suspend the acid (100 mg) in water, aqueous ethanol or ethanol (0.5 ml). Add one drop of phenolphthalein solution followed by sufficient dilute sodium hydroxide, dropwise, to turn the mixture pink (and dissolve the acid if it was not completely dissolved originally). Then add just enough very dilute hydrochloric acid (a drop or two) to completely

decolorize the solution. Add 100 mg of the reagent (p-bromophenacyl bromide, p-phenylphenacyl bromide, p-nitrobenzyl
bromide) dissolved in one ml of ethanol and heat the mixture
under reflux for one hour. Add more ethanol if necessary to
keep the hot mixture homogeneous. Cool, separate the product
which crystallizes (rub with a glass rod if necessary to induce
crystallization), and recrystallize from aqueous ethanol or
methanol.

4. S-Benzyl Isothiouronium Salts

$$R\text{-}COO^- + C_6H_5\text{-}CH_2\text{-}S\text{-}\overset{\overset{+}{NH_2}}{\underset{NH_2}{C}} \longrightarrow R\text{-}COO^- \overset{\overset{+}{NH_2}}{\underset{NH_2}{C}}\text{-}S\text{-}CH_2\text{-}C_6H_5$$

Prepare a neutral solution of the sodium salt of the acid
(100 mg) in water (about 3 ml) as above. Add to this a solution of S-benzyl isothiouronium chloride (0.4 g) in water
(2 ml). Cool the mixture in ice until crystallization is complete. Crystallize the derivative from water or aqueous alcohol.

NOTE: If any S-benzylisothiouronium solution or derivatives
become alkaline, the exceedingly foul smelling benzyl mercaptan
will be formed.

5. Amides and Substituted Amides

(a) Preparation of the Acid Chloride

$$R\text{-}\overset{O}{\overset{\|}{C}}\text{-}OH + SOCl_2 \longrightarrow R\text{-}\overset{O}{\overset{\|}{C}}\text{-}Cl + SO_2 + HCl$$

5. SELECTION AND PREPARATION OF SOLID DERIVATIVES

Convert the acid (100-200 mg) into the acid chloride by heating it under reflux (water condenser) with <u>redistilled</u> thionyl chloride (1 ml) for 30 minutes. Remove the condenser and drive off excess thionyl chloride <u>in the hood</u> by heating the reaction vessel in an oil bath at 100° (if a water bath is used, there is considerable risk of the acid chloride being hydrolysed by water vapor). The last traces of thionyl chloride may be removed, if necessary, by connecting the flask to a water aspirator and warming it gently. This procedure can not be used for lower aliphatic acids since their chlorides are volatile.

(b) <u>Preparation of the Amide</u>.

$$R\text{-}\overset{\overset{\displaystyle O}{\|}}{C}\text{-}Cl + 2NH_3 \longrightarrow R\text{-}\overset{\overset{\displaystyle O}{\|}}{C}\text{-}NH_2 + NH_4Cl$$

Add 2-3 drops of the crude acid chloride from a <u>dry</u> pipet to 2-3 ml of concentrated ammonia, shaking vigorously. Warm the mixture briefly and then cool. If no solid separates, evaporate the solution to dryness. Recrystallize from water or, if the amide is insufficiently soluble in water, wash it with water to remove ammonium salts and then crystallize from aqueous alcohol.

(c) <u>Preparation of the Anilide or p-Toluidide</u>.

$$R\text{-}\overset{\overset{\displaystyle O}{\|}}{C}\text{-}Cl + 2\ C_6H_5\text{-}NH_2 \longrightarrow R\text{-}\overset{\overset{\displaystyle O}{\|}}{C}\text{-}NH\text{-}C_6H_5 + C_6H_5\text{-}NH_3^+Cl^-$$

Add 2-3 drops of the crude acid chloride to a solution of aniline or p-toluidine (5-6 drops or 200 mg) in dry ether (2 ml). Shake the mixture and allow it to stand for ten minutes or until the odor of the acid chloride has disappeared. Add 5 ml of dilute hydrochloric acid, shake, allow to settle, and pipet off the upper (ether) layer into a small beaker or test tube. (If the derivative should separate at this stage, isolate it and later combine it with the product from the ether.) Evaporate the ether layer to dryness (hot water bath) and crystallize the residual anilide or p-toluidide from water, aqueous ethanol or methanol, or (after drying) from a mixture of benzene and petroleum ether b.p. 60-80°.

Alternative procedure: With aromatic acid chlorides, the Schotten-Baumann or pyridine procedures described for preparation of derivatives of amines may also be used.

DERIVATIVES OF PHENOLS

1. Aryloxyacetic Acids

These derivatives crystallize well and are well worth the time invested in their preparation. Furthermore, the equivalent weight (carboxylic acids, section 1) of the derivative may be determined by titration giving a measure of the molecular weight of the unknown phenol.

$$Ar-O^- + Cl-CH_2-COO^- \longrightarrow Ar-O-CH_2-COO^- \xrightarrow{\text{acid}} Ar-O-CH_2-COOH$$

5. SELECTION AND PREPARATION OF SOLID DERIVATIVES

Heat in a boiling water bath for one hour a mixture of the phenol (100 mg), 20% sodium hydroxide (1 ml) and chloroacetic acid (150 mg). Ensure that the mixture remains strongly alkaline; add more sodium hydroxide if necessary. Cool, and acidify the mixture with 5-10% hydrochloric acid. If the solid product separates, isolate and crystallize it (Caution: Sodium chloride may also separate from the solution). If the product does not crystallize directly from the acidified reaction mixture, extract twice with ether (2 ml portions) and evaporate the combined ether layers under reduced pressure. Crystallize the aryloxyacetic acid from hot water, aqueous ethanol or methanol.

2. p-Nitrobenzyl Ethers

$$Ar-O^- + Br-CH_2-C_6H_5-NO_2 \longrightarrow Ar-O-CH_2-C_6H_5-NO_2 + Br^-$$

Heat under reflux for one hour a mixture of the phenol (100 mg), p-nitrobenzyl bromide (100 mg), 20% sodium hydroxide (0.25 ml) and just sufficient alcohol to produce initially a clear solution (a few drops should suffice). Cool the mixture somewhat and add sufficient warm water to dissolve any sodium bromide which may have separated. If the product separates, add hot alcohol until it is almost completely dissolved. Finally cool the solution rapidly with stirring so that it deposits the product as small crystals. If the suspension is not faintly alkaline to phenolphthalein, make it so by adding a few

drops of dilute sodium hydroxide solution. Isolate the p-
nitrobenzyl ether and crystallize it from ethanol, methanol or
aqueous alcohol.

3. Bromination

Add a solution of bromine (0.5 g, 6 drops) in glacial
acetic acid (2 ml) dropwise to a cold solution of the phenol
(100 mg) in the same solvent (1 ml) until the color of bromine
persists. An excess of bromine must be present. Dilute the
solution by dropwise addition of an equal volume of water and
destroy excess bromine with a few drops of sodium sulfite
solution. Crystallize the precipitated product from methanol,
ethanol, acetic acid or a mixture of one of these solvents
with water.

4. p-Nitrobenzoates, 3,5-Dinitrobenzoates and p-Toluenesulfonates

5. SELECTION AND PREPARATION OF SOLID DERIVATIVES

Heat a mixture of the acid chloride (250 mg), the phenol (100 mg) and pyridine (1 ml) at about 100° for 30-60 minutes. Pour the mixture into 3-4 ml of cold water, stir thoroughly and remove the supernatant aqueous layer. Stir the crude product thoroughly with aqueous sodium carbonate to remove any nitrobenzoic acid or unreacted phenol. Crystallize the crude product (which may be an oil) from methanol, ethanol or aqueous alcohol.

Both p-nitrobenzoyl chloride and particularly 3,5-dinitrobenzoyl chloride, especially from opened bottles, often contain so much of the corresponding acids (due to hydrolysis by atmospheric moisture) that they are almost useless. In such cases, the reagent should be regenerated by heating under reflux with thionyl chloride before use, excess thionyl chloride being removed by evaporation under reduced pressure.

5.   Benzoates and p-Toluenesulfonates

$$Ar-O^- + Cl-\overset{\overset{\displaystyle O}{\|}}{C}-C_6H_5 \longrightarrow Ar-O-\overset{\overset{\displaystyle O}{\|}}{C}-C_6H_5 + Cl^-$$

These derivatives are produced by the Schotten-Baumann procedure as detailed under derivatives of amines, section 1. They are crystallized from ethanol or methanol. Benzoates of phenols often have inconveniently low melting points, although that of phenol itself is an excellent derivative. Poor yields of p-toluenesulfonates are often obtained and the pyridine procedure above usually is better.

DERIVATIVES OF ALDEHYDES AND KETONES

1. 2,4-Dinitrophenylhydrazones

$$R_2C=O + H_2N-NH-C_6H_3(NO_2)_2 \longrightarrow R_2C=N-NH-C_6H_3(NO_2)_2 + H_2O$$

The product formed in the diagnostic test for carbonyl
compounds should be sufficient for crystallization to purity.
If a larger quantity is required prepare it in the same way
from 100 mg (3-4 drops) of the aldehyde or ketone and 20 ml of
the reagent.

Crystallize from ethanol, methanol, aqueous alcohol,
alcohol-ethyl acetate, alcohol-chloroform, or glacial acetic
acid.

2. p-Nitrophenylhydrazones

$$R_2C=O + NH_2-NH-C_6H_4-NO_2 \longrightarrow R_2C=N=NH-C_6H_4-NO_2 + H_2O$$

Heat under reflux for ten minutes a mixture of p-nitrophenylhydrazine (100 mg), ethanol (2-3 ml), glacial acetic acid
(1 drop) and the carbonyl compound. Add more ethanol, as
necessary, if the boiling mixture is not homogeneous. The
product crystallizes on cooling. Recrystallize it from methanol, ethanol, glacial acetic acid, or one of these diluted with
water.

3. Semicarbazones and Phenylhydrazones

$$R_2C=O + NH_2-NH-CO-NH_2 \longrightarrow R_2C=N-NH-CO-NH_2 + H_2O$$
$$R_2C=O + NH_2-NH-C_6H_5 \longrightarrow R_2C=N-NH-C_6H_5 + H_2O$$

5. SELECTION AND PREPARATION OF SOLID DERIVATIVES

Shake together the carbonyl compound (100 mg), water (2 ml), semicarbazide hydrochloride or phenylhydrazine hydrochloride (200 mg) and sodium acetate (300 mg). If the carbonyl compound is very insoluble in water, add a little ethanol also. If no solid derivative starts to separate within five minutes, heat the mixture in a hot water bath, with shaking, for 5-15 minutes. The derivative should crystallize upon cooling in ice and rubbing with a glass rod. Recrystallize it from methanol, ethanol or aqueous alcohol.

4. **Dimethones and Dimethone Anhydrides (Derivatives of aldehydes only)**

$$R-CHO + 2$$

Dimethone   Dimethone Anhydride

If the aldehyde is soluble or moderately soluble in water, use a cold saturated solution of the reagent (dimedone, 5,5-dimethylcyclohexane-1,3-dione) in water. Otherwise, use a 10% solution in ethanol. Mix the aldehyde (100 mg) with saturated aqueous (10 ml) or 10% alcoholic (2 ml) dimedone solution. Separation of the solid may commence immediately, or after several minutes or some hours, depending on the reactivity of the aldehyde. Cool the mixture in ice to complete crystallization and rub with a glass rod if necessary. Recrystallize the dimethone from <u>dilute</u> aqueous ethanol or methanol. Do not boil it or heat it for long periods with alcohol alone, since partial conversion into the anhydride may occur. The dimethone may be converted into the dimethone anhydride by heating it under reflux with glacial acetic acid or acetic acid containing some acetic anhydride. Dimethone anhydrides are insoluble in dilute sodium hydroxide and may thus be separated from unchanged dimethones which, being enols, are soluble.

5. <u>Oximes</u>

$$R_2C=O + H_2N-OH \longrightarrow R_2C=N-OH + H_2O$$

Mix the carbonyl compound (100 mg) with hydroxylamine hydrochloride (250 mg) and 5% sodium hydroxide (2 ml). If the carbonyl compound is very insoluble in water, add a little ethanol to clarify the solution. Heat the mixture under reflux for 10-15 minutes. Cool in ice and rub with a glass rod if necessary to induce crystallization. Recrystallize the product from methanol, ethanol, aqueous alcohol, or (after drying) benzene or petroleum ether b.p. 60-80°.

DERIVATIVES OF ESTERS

Rarely is it practicable to prepare derivatives of a complete ester molecule, except in cases where the ester contains other functional groups which can be utilized independently of the ester function (e.g. acetylsalicylic acid, methyl

5. SELECTION AND PREPARATION OF SOLID DERIVATIVES

salicylate, ethyl hydrogen phthalate). Usually, the ester is hydrolyzed or otherwise separated into its component acid and alcohol (or derivatives of these) which are identified separately.

A valuable characteristic of the intact ester is its saponification equivalent, that weight of the ester which, upon complete hydrolysis, gives sufficient acid to neutralize one equivalent of a base. This should always be determined. The saponification equivalent of the ester is equal to the molecular weight or a simple fraction thereof. Usually this value, together with an identification of crystalline derivatives of the acid component and some physical properties (boiling point, refractive index and/or density) will identify the ester beyond reasonable doubt. Occasionally, particularly in advanced work, it is necessary to identify the alcohol component via crystalline derivatives.

1. Determination of Saponification Equivalent and Isolation of the Acid

$$R-CO-O-R' + OH^- \longrightarrow R-COO^- + R'-OH$$

Weigh accurately (analytical balance) a clean, dry, 50-100 ml, standard-taper flask (with stopper if the ester to be hydrolyzed is volatile). Place about 200 mg of the ester in the flask, stopper if necessary, and reweigh accurately. The difference in weight is the weight of ester used. Add to the

ester 20 ml of neutral ethanol (if necessary, add to the
ethanol one drop of phenolphthalein indicator and the absolute
minimum of sodium hydroxide solution to produce the faintest
pink color) and 10 ml (accurately measured with a 10 ml volumetric pipet) of approximately 0.5 normal sodium or potassium
hydroxide solution (this solution is prepared by dissolving
about 2 g of solid sodium hydroxide or 2.8 g of potassium
hydroxide in about 100 ml of water; it should be freshly prepared and kept stoppered to minimize absorption of atmospheric
carbon dioxide). Attach a clean water-cooled standard-taper
condenser to the flask and heat the mixture under gentle reflux
for 1.5 to 2 hours. The mixture should be homogeneous. Cool
the mixture, add two drops of phenolphthalein indicator and
titrate the solution with standard hydrochloric acid (0.1 or
0.2 normal) until just colorless. Separately, standardize
the alkali by titrating a separate 10 ml portion against the
same standard acid, again using phenolphthalein as indicator.

If W is the weight of ester used (milligrams)

    v is the volume of acid used to back-titrate uncon-
       sumed alkali in the saponification mixture (ml)

    V is the volume of acid used to titrate the separate
       10 ml portion of alkali (ml)

    N is the normality of the acid

Saponification equivalent = $\dfrac{W}{N(V-v)}$

5. SELECTION AND PREPARATION OF SOLID DERIVATIVES

The acid, if solid, is conveniently isolated from the saponification mixture and constitutes a useful derivative of the ester. Evaporate the neutralized saponification solution to dryness (an oil bath at 120-130° is convenient) and acidify the residual potassium salt of the acid with dilute hydrochloric acid. A solid acid will crystallize and may be recrystallized from water or aqueous alcohol. A volatile liquid acid may be recognized by its smell; such an acid can be isolated but the procedure is tedious.

2. Refractive Index

Details of the instrument and its use are given in chapter 3. Consult the instructor before using the refractometer.

3. Identification of the Acid Component

(a) As p-Nitrobenzyl Ester, p-Bromphenacyl Ester, p-Phenylphenacyl Ester or S-Benzyl Isothiouronium Salt.

Dissolve about 100 mg (one pellet) of potassium hydroxide in two drops of warm water. Add one ml of ethanol and 100 mg (3 drops) of the ester. Stopper the test tube loosely and heat it in a boiling water bath for 5-10 minutes. Cool the solution, add one drop of phenolphthalein followed by dilute hydrochloric acid, dropwise, until the pink color just disappears. Prepare the derivatives from the resulting solution of the potassium salt of the acid (alternatively, evaporate the neutral solution from the determination of the saponification equivalent to dryness in an oil bath at 120-130° and use the residual potassium salt) as described under carboxylic acids, section 2.

(b) As Anilide or p-Toluidide

$$R-CO-O-R' + H_2N-Ar \xrightarrow{\text{OMe}^-} R-CO-NH-Ar + R'-OH$$

Dissolve a small pellet of sodium (20-50 mg) in one ml of dry methanol (super-dry methanol is desirable but ordinary methanol is satisfactory as long as it has not been unduly exposed to atmospheric moisture). Add to the solution 100 mg of the ester and 100 mg of dry aniline or p-toluidine. Heat the mixture under reflux (preferably in an oil bath) for 30 minutes. The anilide or p-toluidide usually crystallizes on cooling. Addition of a little water may be desirable to complete the crystallization. Recrystallize the derivative from aqueous alcohol, methanol or ethanol.

4. Identification of the Alcohol Component

It is rarely necessary to identify the alcohol component of a simple ester if the saponification equivalent and refractive index are known and the acid component is adequately characterized. The following procedure is fairly widely applicable and usually is simpler than procedures necessitating isolation of the free alcohol prior to its conversion into derivatives.

Heat, at the reflux temperature or 150°, whichever is lower, a mixture of the ester (0.2 g), concentrated sulfuric acid

183

5. SELECTION AND PREPARATION OF SOLID DERIVATIVES

(1 small drop) and 3,5-dinitrobenzoic acid (0.2 g). If the mixture becomes homogeneous within 15 minutes, heat it for 30 minutes; otherwise heat it for one hour. Extract the reaction mixture twice with 5 ml portions of sodium carbonate and discard the aqueous sodium carbonate layers. Dissolve the product, which usually is an oil, in 1 ml of hot alcohol, add hot water, dropwise, until the product starts to separate, and initiate crystallization by cooling and rubbing with a glass rod.

DERIVATIVES OF ALCOHOLS

In principle, the preparation of derivatives of alcohols is a simple matter. In practice, however, for a variety of purely practical reasons, it is often quite tricky and should be attempted only after careful consideration of the probable nature of the particular alcohol in hand (as revealed by its b.p. and other simple observations) and with a thorough appreciation of the difficulties characteristic of each procedure.

Thus, complex alcohols (cholesterol, glycerol, glucose e.g.) are readily converted into crystalline benzoates and acetates. However, the corresponding esters of simple alcohols are liquids and those with more complex acids (p-nitrobenzoates, 3,5-dinitrobenzoates), while solids, are less easily prepared, are difficult to separate from unreacted acid chloride or acid, and show a marked tendency to separate as oils upon attempted crystallization. Half-esters of simple alcohols with

3-nitrophthalic acid are readily prepared and crystallize well but many of them have closely similar melting points and so do not afford clear differentiation. Urethanes (especially α-naphthyl urethanes) are excellent derivatives of simple alcohols. However, if the alcohol contains traces of water this reacts with the reagent to give the corresponding urea which is difficult to separate from the derivative of the alcohol.

Most secondary alcohols are readily oxidized to ketones. These are very conveniently characterized as their 2,4-dinitrophenylhydrazones. Primary alcohols, similarly, are oxidized to aldehydes; these, however, tend to oxidize further to carboxylic acids unless excess of the alcohol is used. If strong acid is present, the primary alcohol may rearrange to a secondary alcohol which then is oxidized to a ketone. Thus, oxidation of n-propyl alcohol with dichromate in sulfuric acid can give more acetone than propionaldehyde! The risk of rearrangement is considerably reduced if a less acidic oxidizing agent such as chromium trioxide in glacial acetic acid is used.

Tertiary alcohols present a special problem. They are not readily oxidized to simple products. Their reactivity is often low, owing to steric hindrance and they are readily dehydrated to alkenes, especially by acidic reagents such as acid chlorides.

5. SELECTION AND PREPARATION OF SOLID DERIVATIVES

1. Refractive Index

Simple alcohols are non-corrosive liquids. Their refractive indexes may be determined and afford a valuable characterization.

Details of the refractometer and its use are given in chapter 3. Read these carefully, and consult the instructor before proceeding with the determination. The sample must be liquid and must be pure. If it is suspected to contain water or other impurities, it should be dried over anhydrous magnesium or sodium sulfate, distilled, and a middle-cut of the distillate used, the forerun and after-run being kept for other purposes such as conversion into solid derivatives.

2. Oxidation of the Alcohol and Characterization of the Ketone or Aldehyde (secondary and most primary alcohols only)

Dissolve the alcohol (6-8 drops) in glacial acetic acid (1 ml). Add, with shaking, to the cold solution a solution of chromium trioxide (50 mg) in two drops of water (do not use excess CrO_3 or this will oxidise an aldehyde product to carboxylic acid and also may oxidise methanol in the dinitrophenylhydrazine reagent used later to formaldehyde). If the orange-red color of the CrO_3 is not completely discharged and replaced by the green of chromic ion within a few minutes, warm the mixture in a water bath. Do not heat it above the b.p. of

the expected carbonyl compound. The characteristic smell of a
volatile aldehyde (acetaldehyde, propionaldehyde, n-butyalde-
hyde, benzaldehyde, vanillin) is often evident at this stage,
especially if the masking odor of acetic acid is reduced by
diluting 2-3 drops of the reaction mixture with water. Add
the cooled reaction mixture dropwise to 2,4-dinitrophenyl-
hydrazine reagent (4 ml), warm to 50-60° if necessary to complete formation of the derivative, cool, isolate, and recrystallize the dinitrophenylhydrazone.

3. 3,5-Dinitrobenzoates and p-Nitrobenzoates

$$R\text{-}OH + Cl\text{-}CO\text{-}C_6H_3(NO_2)_2 \longrightarrow R\text{-}O\text{-}CO\text{-}C_6H_3(NO_2)_2 + HCl$$

If good 3,5-dinitrobenzoyl chloride or p-nitrobenzoyl
chloride is available it may be used directly. Otherwise,
generate it as follows: Heat under reflux for 30 minutes the
appropriate acid (or part-hydrolyzed chloride) (0.2 g) and
thionyl chloride (2 ml) (an oil bath is preferable to a water
bath). Remove the condenser and boil off excess thionyl chloride by heating the mixture at 100° (oil bath; hood). Add the
alcohol (0.2 to 0.5 ml) and heat at the reflux temperature or
at 100° (whichever is lower) for 10-15 minutes. Cool, add 5 ml
of 5% sodium carbonate solution and agitate the mixture
thoroughly, rubbing with a glass rod until the product solidifies. Separate and wash it with water. Dissolve it in the
minimum of hot alcohol, add hot water, drop by drop, until a

slight permanent turbidity develops, and cool slowly, with rubbing. The 3,5-dinitrobenzoates of simple alcohols show a marked tendency to separate as oils upon attempted crystallization.

4. Half-esters of 3-Nitrophthalic Acid

3-Nitrophthalic anhydride is readily hydrolyzed to the acid and should be regenerated before use. Heat the crude anhydride (or the acid) (0.2 g) at 200° (oil bath) for 5-10 minutes. Cool, add the alcohol (0.2-0.3 ml) and heat at the boiling point of the alcohol or 100° until the solid anhydride is completely dissolved and for a few minutes more. Cool and rub with a glass rod, adding a few drops of water if necessary to induce crystallization. Recrystallize from hot water, aqueous alcohol, benzene or petroleum ether.

DERIVATIVES OF ETHERS

The ether function is unreactive and does not undergo reactions giving convenient crystalline derivatives. The ether linkage can be split, for example, by concentrated hydriodic acid, concentrated phosphoric acid or anhydrous aluminum chloride but the products vary with the reagent, the reaction conditions and the nature of the particular ether. Details will be found in more advanced manuals.

$$CH_3-O-CH_2-CH_3 \xrightarrow{HI} CH_3-I + I-CH_2-CH_3 + H_2O$$

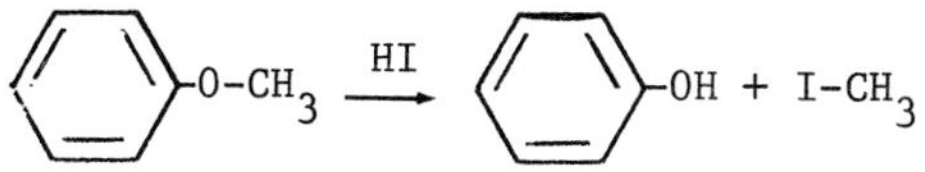

Identification rests, therefore, upon careful determination of the physical properties of the pure (redistilled) ether, in particular the boiling point, the refractive index, and the density. Benzenoid ethers may be converted into solid derivatives by substitution (nitration, bromination, chlorosulfonation) on the ring. More complex benzenoid ethers (at least two rings) form crystalline complexes with picric acid.

1. Refractive Index

Details of the refractometer and its use are given in chapter 3.

2. Relative Density

The relative density may be determined by the capillary tube method or, with greater accuracy, by means of a pycnometer. Procedures are given in chapter 3.

3. Nitration and Bromination (Aryl Ethers only)

Procedures are given in the following section, derivatives of benzenoid hydrocarbons and aryl halides. The probable identity of the ether as revealed by its boiling point, refractive index, and density must be known so that the appropriate procedure, indicated in the table of ethers in chapter 6, may be selected.

4. Sulfonamides (Aryl Ethers only)

Add chlorosulfonic acid (1 ml; caution: extremely corrosive) to an ice-cooled solution of the sample (0.3 g) in chloroform (1.5 ml). When reaction subsides, warm the mixture to room temperature for 20-30 minutes, add ammonium carbonate (1 g) cautiously, mix thoroughly, and evaporate the mixture to dryness at 100°. Extract out inorganic salts with water and recrystallize the sulfonamide from aqueous alcohol.

5. Picrates (Complex Aromatic Ethers Containing at Least Two Benzene Rings, a Naphthalene Ring, etc.

Dissolve picric acid (0.2 g) in the minimum of hot ethanol. Add the ether (0.15 g) to the hot solution. If the product does not crystallize out on cooling, pour the solution into a 100 ml beaker and allow the solution to evaporate at room temperature until it does. Recrystallize the picrate from the

190

minimum of chloroform or ethanol. Determine the melting point immediately and as rapidly as possible.

Alternative procedure: Melt together (below 120°) picric acid (0.2 g) and the sample (0.15 g). If crystallization does not occur spontaneously upon cooling, rub the mixture with a glass rod and triturate it with a few drops of ethanol or chloroform. Recrystallize as before.

Caution: if the product melts at about 120°, it may be recovered picric acid rather than the picrate.

6. Side-chain Oxidation to Carboxylic Acids (Aryl Ethers only)

Benzenoid ethers having alkyl side chains on the benzene ring may be oxidized to the corresponding carboxylic acids.

Procedural details are given in the following section, derivatives of benzenoid hydrocarbons. Melting points of the relevant alkoxybenzoic acids will be found in the table of benzenoid carboxylic acids.

DERIVATIVES OF BENZENOID HYDROCARBONS AND ARYL HALIDES

Benzenoid (aromatic) hydrocarbons and halogenohydrocarbons may be identified by their physical properties (m.p., b.p., refractive index, density), by derivatives resulting from

5. SELECTION AND PREPARATION OF SOLID DERIVATIVES

ring-substitution (nitration, bromination, chlorosulfonation, Friedel-Crafts acylation) and by side-chain oxidation to carboxylic acids. In addition, polycyclic hydrocarbons form crystalline complexes with picric acid. Activated aryl halides (e.g. 2,4-dinitrochlorobenzene) react with nucleophiles (thiourea) as do alkyl halides.

1. Refractive Index and Density

Procedures for these determinations are given in chapter 3. Refractive indexes should not be determined for highly reactive aryl halides.

2. Sulfonamides

$$Ar-H + 2\ Cl-SO_3H \longrightarrow Ar-SO_2-Cl \xrightarrow{NH_3} Ar-SO_2-NH_2$$

Procedural details are as given for phenol ethers in the preceding section. Hydrocarbons, and especially aryl halides, are less reactive than phenol ethers and so should be heated to 40-60° with chlorosulfonic acid to complete the substitution reaction.

3. Aroylbenzoic Acids

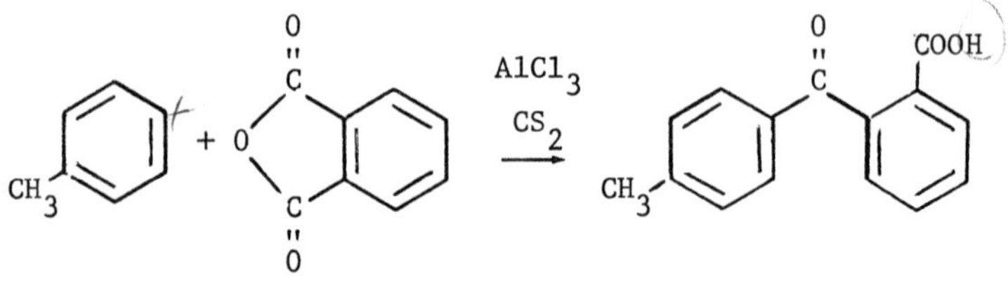

DERIVATIVES OF BENZENOID HYDROCARBONS AND ARYL HALIDES

Dehydrate phthalic anhydride (0.25 g) by heating it at 200°
for several minutes. Dissolve the solid anhydride in carbon
disulfide (2-4 ml; care: foul smell, highly flammable), add
the sample (0.2 g) and anhydrous aluminum chloride (0.5 g), and
heat the mixture under reflux for 30 minutes in a fume hood.
Cool, dilute the mixture with dilute hydrochloric acid (5 ml),
drive off the carbon disulfide by warming the mixture (water-
bath, hood, no flames) and isolate the crude solid aroyl benzoic acid. If the product is very badly discolored, dissolve
it in 5-10 ml of hot dilute ammonia, boil the solution for a
few minutes with a little decolorizing carbon (Norit), filter
it hot to remove the carbon and precipitate the acid by acidi-
fying the filtrate. Recrystallize the aroylbenzoic acid from
aqueous alcohol.

4. Side-chain Oxidation (Substances Having Alkyl Side-chains
   only)

$$Ar-R \xrightarrow{KMnO_4} Ar-COOH$$

Add potassium permanganate (1 g) slowly to a heated (100°;
boiling water bath) mixture of sample (0.25 g), and 5% sodium
hydroxide or sodium carbonate solution (3 ml). Heat the mixture under reflux until the purple color of permanganate has
disappeared (1-4 hours). Cool the mixture, filter off manganese dioxide, acidify the filtrate with dilute (10%) sulfuric
acid, and destroy any colloidal manganese dioxide (indicated by

a brown solution) with a _little_ sodium sulfite or bisulfite.
If the acid precipitates or crystallizes on cooling, collect
it. Otherwise, extract the solution with ether or carbon
tetrachloride and recover the acid by evaporating the solvent
layer to dryness. Recrystallize it from water.

5. Nitration

Procedures vary with the reactivity of the ring to electrophilic substitution so that some indication of the probable
identity of the unknown substance is necessary. The appropriate procedure is indicated by code letter (a-e) in the tables,
chapter 6, for many of the hydrocarbons, aryl halides and
ethers listed.

(a) Mix carefully and cool to room temperature 1 ml of
fuming nitric acid and 2 ml of concentrated sulfuric acid. Add
the sample (0.2 ml), shake the mixture, and, when any initial
reaction has subsided, heat it at 100° (water bath) for 2-5
minutes. Pour the mixture into 10 ml of water; rub the oily
product with a glass rod until it solidifies.

(b) Proceed as above (a), but use 2 ml each of concentrated (_not_ fuming) nitric and sulfuric acids and warm the mixture only to 40-50° until reaction is complete.

(c) Add fuming nitric acid slowly, dropwise, to the sample
(0.2 ml) cooled in ice. Warm the mixture to room temperature
for five minutes and precipitate the product by pouring it into
water.

(d)   Add a mixture of fuming nitric acid (0.5 ml) and glacial acetic acid (0.5 ml) to the substance (0.2 ml) in acetic acid (0.5 ml). Heat the mixture to boiling, cool, and precipitate the product by diluting the mixture with water.

(e) Procedure as (d) but without heating.

6. Bromination

As with nitration, the procedure varies with structure. Appropriate procedures are indicated in chapter 6.

(f) Treat the substance with bromine at room temperature until the red color persists. After ten minutes, destroy excess bromine with sodium carbonate solution and recrystallize the solid product.

(g) Procedure as (a) but with a catalytic trace of iron powder (filings) or iodine, heating if necessary to initiate reaction, and finally at 70° for several minutes.

(h) Procedure as (a), but with both the sample and the bromine dissolved in glacial acetic acid.

(i) Procedure as (a) but with the sample and the bromine dissolved in carbon tetrachloride.

(j) Procedure as (d) but with ether in place of CCl_4.

7. Picric Acid, Styphnic Acid and Trinitrobenzene Complexes (Polycyclic Compounds only)

See preceding section (ethers) for details.

5. SELECTION AND PREPARATION OF SOLID DERIVATIVES

DERIVATIVES OF ALKYL HALIDES

Alkyl halides may be identified by their physical properties (m.p., b.p., refractive index, density) and by means of derivatives resulting from nucleophilic substitution of the halogen atom.

1. Refractive Index and Density

Details are given in chapter 3. Since many alkyl halides contain traces of free halogen and halogen hydrides, which are corrosive, the refractometer should be very thoroughly cleaned with cotton soaked in acetone immediately after use.

2. S-Alkyl Isothiouronium Picrates

$$R-X + HS-C(=NH)-NH_2 \longrightarrow R-S-C(=NH_2^+)-NH_2 \; \overline{X}$$

$$\longrightarrow R-S-C(=NH_2^+)-NH_2 \; \overline{O}-\!\!\left\langle\!\!\!\begin{array}{c} NO_2 \\ \\ NO_2 \end{array}\!\!\!\right\rangle\!\!-NO_2$$

Heat under reflux a mixture of the alkyl halide (0.1 ml), thiourea (0.1 g) and ethanol (1 ml) for 5-30 minutes. If the halide is a chloride, include a crystal of potassium iodide in the reaction mixture. Add picric acid (0.1 g) to the hot reaction mixture, heat further, and add a little more alcohol if necessary to obtain a clear solution. Cool, isolate the product which crystallizes, and recrystallize it from alcohol.

196

ALKANES, ALKENES AND ALKYNES

Aliphatic hydrocarbons (saturated and unsaturated) do not undergo reactions which lend themselves to the convenient preparation of crystalline derivatives. They are identified by their physical properties (m.p., b.p., refractive index, density) considered together with the results of qualitative tests (unsaturation) and the demonstrated absence of all other functional groups.

6

TABLES OF PHYSICAL PROPERTIES

"Sea water is CH_2O."

"Sparrows and starlings eat the farmer's grain and soil his corpse."

"Sound is a rapid series of osculations."

The melting points, boiling points, and in appropriate instances the refractive indexes and densities of some typical and common members of the functional-group classes are tabulated in this chapter, together with the melting points of some of their derivatives. The lists are far from comprehensive but should be adequate for introductory work. More comprehensive tables are to be found in the following works and elsewhere:

"Practical Organic Chemistry" by A. I. Vogel. Wiley,
New York, 1966.

"Organic Reagents for Organic Analysis." Hopkin and
Williams Ltd., Chadwell Heath, Essex, England,
1950.

"The Systematic Identification of Organic Compounds"
by R. L. Shriner, R. C. Fuson and D. Y. Curtin.
Wiley, New York, 1964.

199

6. TABLES OF PHYSICAL PROPERTIES

"Handbook of Tables for the Identification of Organic
Compounds" by V. Rappoport. The Chemical Rubber
Co., 1967.

"Dictionary of Organic Compounds" (earlier "Heilbron's
Dictionary of Organic Compounds"). Eyre and
Spottiswoode, London, 1965, 5 volumes.

Several of these works offer deeper study of the subject,
consider additional functional groups, and describe alternative diagnostic and confirmatory tests, additional techniques
and some completely different approaches to the whole field.

The identification of derivatives prepared is not the sole
use of these tables. Judicious use of them will indicate which
derivatives should be prepared so that the maximum of information may be gained from the minimum of effort.

There are many blank spaces in the tables. This does not
necessarily mean that the corresponding derivatives do not
exist, cannot be prepared, or are liquids. It simply means
that they are not recorded here and reflects the fact that all
compendia, no matter how comprehensive, are only as complete
as the author wishes to make them.

Transcription errors are almost inevitable in tables containing masses of numerical data. As much of the information
as possible has been checked with original literature and at
least two of the secondary sources cited. In cases of major

6. TABLES OF PHYSICAL PROPERTIES

irreconcilable divergence, the item has been eliminated. The
author would appreciate notification of any remaining discre-
pancies.

ŋ

ALIPHATIC PRIMARY AMINES

	B.P.	p-Toluenesulfonamide	Benzenesulfonamide	Phthalimide	Picrate
Methylamine	-7	77	30	134	215
Ethylamine	19	63	58	78	168
n-Propylamine	48	52	36	66	135
iso-Propylamine	35	51	98	86	150
Allylamine	55	64	39	70	140
n-Butylamine	77			34	151
iso-Butylamine	68	78	53	93	151
sec-Butylamine	64	55	70		140
n-Hexylamine	129		96		127
cyclo-Hexylamine	134	87	89	168	158
n-Heptylamine	155			40	121
Benzylamine	185	116	88	115	198
2-Phenylethylamine	198	64	69	130	170
Ethanolamine	171			127	160
Ethylenediamine	117	160	168		233
Propylenediamine	120				137

AROMATIC PRIMARY AMINES

	B.P.	M.P.	p-Toluenesulfonamide	Benzenesulfonamide	Benzamide	Acetamide	Formamide	Picrate
Aniline	183		103	112	163	114	47	165
o-Toluidine	200		110	124	144	109	59	200
m-Toluidine	203		114	83	125	66		195
p-Toluidine	200	44	118	120	158	145	53	171
o-Anisidine	225		127	89	60	88	83	200
m-Anisidine	251		68			80	57	169
p-Anisidine	246	57	114	96	154	127	81	
o-Phenetidine	228		164	102	104	79	62	
m-Phenetidine	248		157		103	96	52	158
p-Phenetidine	254		107	143	173	135	69	
o-Chloroaniline	209		105	130	99	88	80	134
m-Chloroaniline	230		138	121	122	73	58	177
p-Chloroaniline	232	71	95	122	193	179	102	178
o-Bromoaniline	229	32	90		116	99	89	129
m-Bromoaniline	251	18			135	88	63	180
p-Bromoaniline		66	101	134	204	167	119	180
o-Iodoaniline		58			139	110		112
m-Iodoaniline		25	128		151	119		
p-Iodoaniline		63			212	181	109	

AROMATIC PRIMARY AMINES
(Continued)

	B.P.	M.P.	p-Toluene-sulfonamide	Benzene-sulfonamide	Benzamide	Acetamide	Formamide	Picrate
o-Nitroaniline		71	110	104	98	94	122	73
m-Nitroaniline		114	139	136	157	155	134	143
p-Nitroaniline		148	191	139	199	216	194	100
2,4-Dichloroaniline	245	63	126	128	117	146	153	106
2,5-Dichloroaniline	251	50			120	132	147	
2,4-Dibromoaniline		79	134		134	146	146	124
2,4-Dinitroaniline		180	219		220	121		
2,4,6-Trichloroaniline		78			174	206	180	83
2,4,6-Tribromoaniline		120			198	232	222	
3-Bromo-4-aminotoluene		26			149	117		
5-Bromo-2-aminotoluene		59			115	157		
3-Nitro-4-aminotoluene		117	100	101	143	92	125	
4-Nitro-2-aminotoluene		107		172	183	151	186	
5-Nitro-2-aminotoluene		129	174	158	178	200	162	
o-Aminophenol		174	139	141	184	124	129	
m-Aminophenol		123	157		153	101		
p-Aminophenol		186	253	154	234	150	140	
1-Naphthylamine	300	50	157	169	161	160	139	163
2-Naphthylamine	294	113	133	98	162	134	129	195
o-Aminodiphenyl	299	50			102	121	75	
p-Aminodiphenyl	302	51	255		230	171	172	

ALIPHATIC SECONDARY AND TERTIARY AMINES

	B.P.	p-Toluenesulfonamide	Benzenesulfonamide	Picrate
Dimethylamine	7	79	47	158
Diethylamine	55	60	42	74
Di-n-propylamine	108		51	75
Di-iso-propylamine	84			140
Di-n-butylamine	159			59
Di-iso-butylamine	137		55	119
Piperidine	106	96	94	148
Pyrrolidine	89	123		112
Morpholine	128	147	119	146
Piperazine	140 (m.p.104)	173	282	280
Trimethylamine	3			216
Triethylamine	89			173
Tri-n-propylamine	156			117
Tri-n-butylamine	212			106

6. TABLES OF PHYSICAL PROPERTIES

AROMATIC SECONDARY AMINES

	B.P.	M.P.	p-Toluenesulfonamide	Benzenesulfonamide	Benzamide	Acetamide	Formamide	Picrate
N-Methylaniline	194		95	79	63	103		145
N-Ethylaniline	205		88		60	55		138
N-n-Propylaniline	222		56	54		47		
N-n-Butylaniline	240		54		56			
N-Methyl-o-toluidine	208		120		66	56		90
N-Methyl-m-toluidine	206					66		
N-Methyl-p-toluidine	210		60	64	53	83		131
N-Ethyl-o-toluidine	214		75	62	73			
N-Ethyl-m-toluidine	221				72			
N-Ethyl-p-toluidine	217		71	66	39			
o-Nitro-N-methylaniline		37				70		
m-Nitro-N-methylaniline		68		83	105	95		
p-Nitro-N-methylaniline	152	177	120	111	152	119		
N-Benzylaniline	306	38	140	119	107	58	48	
Diphenylamine	302	54	142	123	180	103	74	182
Dibenzylamine	300		81	68	112		52	
N-Methyl-1-naphthyl-amine	294		164		121	94		
N-Methyl-2-naphthyl-amine	317		78	107	84	51		145

6. TABLES OF PHYSICAL PROPERTIES

AROMATIC TERTIARY AMINES

	B.P.	M.P.	p-Nitroso cpd.	Picrate
Dimethylaniline	193		87	164
Diethylaniline	215		84	142
Methylethylaniline	201			134
Methylbenzylaniline	306			127
Dibenzylaniline		72	91	132
Dimethyl-o-toluidine	185			122
Dimethyl-m-toluidine	212			
Dimethyl-p-toluidine	211			130
Dimethyl-1-naphthylamine	273			145
Dimethyl-2-naphthylamine	305	47		200
p-Nitrodimethylaniline		163		
p-Bromodimethylaniline	264	55		
Triphenylamine	365	127		
Tribenzylamine		92		190
Pyridine	115			167
2-Picoline	129			169
3-Picoline	144			150
4-Picoline	143			167
Quinoline	238			203

ALIPHATIC CARBOXYLIC ACIDS

	B.P.	M.P.	Anilide	p-Toluidide	p-Bromophenacyl ester	p-Phenylphenacyl ester	p-Nitrobenzyl ester	S-B-T-salt	Amide
Formic	101	8	50	53	140	74	31	150	
Acetic	118	16	114	153	86	111	78	135	82
Propionic	141		106	126	59	102	31	151	79
n-Butyric	163		96	75	63	82	35	146	116
iso-Butyric	154		105	109	77	89		143	129
n-Valeric	187		63	74	75	63			63
iso-Valeric	174		113	109	68	78		153	136
n-Hexoic	205		95	74	72	70			100
n-Heptoic	223		65	80	72	62			95
Palmitic		63	91	98	86	94	43	141	106
Stearic		70	94	102	90	97		143	109
Chloroacetic	189	63	137	162	105	116		159	120
Dichloroacetic	194		119	153	99			178	97
Trichloroacetic	196	58	95	113			80	148	141
Bromoacetic	208	50	130	91			89		91
Iodoacetic		83	144						95
Acrylic	140		105	141					85
Crotonic	189	72	118	132	95		67	162	150

ALIPHATIC DICARBOXYLIC ACIDS

	M.P.	Anilide	p-Toluidide	p-Bromophenacyl ester	p-Phenylphenacyl ester	p-Nitrobenzyl ester	S-B-T-salt	Amide
Oxalic	100	246	268	242	165	204	195	
Malonic	135	225	253		175	86	147	170
Succinic	185	229	255	211	208	88	155	260
Glutaric	98	224	218	137	152	69		175
Adipic	152	239	241	155	148	106		220
Pimelic	105	156	206	137	146			
Azelaic	106	187	202	131	141	44		172
Malic	101	197	207	179		124	124	157
Maleic	135	187	142			89	163	181
Fumaric	286	314				151	178	266

AROMATIC CARBOXYLIC ACIDS

	B.P.	M.P.	Anilide	p-Toluidide	p-Bromophenacyl ester	p-Phenylphenacyl ester	p-Nitrobenzyl ester	S-B-T-salt	Amide
Benzoic	249	122	162	158	119	167	89	167	129
o-Toluic	259	105	125	144	57	95	91	146	140
m-Toluic	263	112	125	118	108	136	87	140	95
p-Toluic	274	181	144	160	153	165	104	164	159
o-Chlorobenzoic		141	118	131	107	123	106		141
m-Chlorobenzoic		158	124		117	154	107	155	134
p-Chlorobenzoic		240	194		126	160	130		179
o-Bromobenzoic		150	141		102	98	110	171	155
m-Bromobenzoic		155	146		126	155	105	168	155
p-Bromobenzoic		252	197		134	160	141		189
o-Iodobenzoic		162	141		110	143	111		184
m-Iodobenzoic		187			128		121		186
p-Iodobenzoic		270	210		146	171	141		218
o-Nitrobenzoic		147	155		101	140	112	159	175
m-Nitrobenzoic		141	154	162	137	153	142	163	142
p-Nitrobenzoic		239	211	203	136	182	169	182	200
Phenylacetic	265	76	118	136	89	63	65	190	157
Diphenylacetic		148	180	173		111	104		168

AROMATIC CARBOXYLIC ACIDS
(Continued)

	M.P.	Anilide	p-Toluidide	p-Bromophenacyl ester	p-Phenylphenacyl ester	p-Nitrobenzyl ester	S-B-T-salt	Amide
Salicylic	159	135	156	140	148	98	148	139
m-Hydroxybenzoic	201	157	163	176		108		167
p-Hydroxybenzoic	215	197	208	184	240	192	145	162
o-Methoxybenzoic	101	131		113	131	113		129
Anisic	184	171	186	152	160	132	185	163
Phthalic	208	254		153	167	155	158	220
iso-Phthalic	347			179		203	216	280
Terephthalic	>300	337		225		264	204	
3-Nitrophthalic	219	234	223			190		201
4-Nitrophthalic	165	192			120			200
Cinnamic	133	153	168	146	182	117	179	147
o-Nitrocinnamic	240			142	146	132		185
m-Nitrocinnamic	205			178		174		196
p-Nitrocinnamic	287			191	192	187		217
1-Naphthoic	161	163		135				202
2-Naphthoic	185	170	191					192
Diphenylacetic	148	180	173		111	104		168
Acetylsalicylic	135	136				90		138

6. TABLES OF PHYSICAL PROPERTIES

PHENOLS

	B.P.	M.P.	Aryloxyacetic acid	p-Nitrobenzyl ether	Bromo compound	p-Nitrobenzoate	3,5-Dinitrobenzoate	p-Toluenesulfonate	Benzoate
Phenol	182	43	99	91	95	126	146	96	69
o-Cresol	191	30	152	90	56	94	138	55	
m-Cresol	202		103	51	84	90	165	51	55
p-Cresol	202	34	136	88	49	98	189	70	71
o-Chlorophenol	175		145	100		115	143	74	
m-Chlorophenol	214	33	110			99	156		71
p-Chlorophenol	217	43	156	101		168	186	71	90
o-Bromophenol	194		143	110	95			78	
m-Bromophenol	236	33	108					53	86
p-Bromophenol	235	64	159	114	95	180	191	94	102
o-Iodophenol		43	135					80	34
m-Iodophenol		40	115			133	183	61	73
p-Iodophenol		94	156					99	119
o-Nitrophenol	216	45	158	130	117	141	155	83	59
m-Nitrophenol		97	156		91	174	159	113	95
p-Nitrophenol		114	187	187	142	159	186	97	142
Catechol	240	105				192	169	152	84
Resorcinol	280	110	195		112	182	201	81	117
Hydroquinone	286	170	250		186	250	317	159	205

PHENOLS
(Continued)

	B.P.	M.P.	Aryloxyacetic acid	p-Nitrobenzyl ether	Bromo compound	p-Nitrobenzoate	3,5-Dinitrobenzoate	p-Toluene-sulfonate	Benzoate
2,3-Dimethylphenol	218	75	187			126			
3,4-Dimethylphenol	225	62	163			171	181		58
2,4-Dimethylphenol	211	28	142			105	164		38
3,5-Dimethylphenol	219	68	86	166		109	195	83	
2,5-Dimethylphenol	211	75	118	178	87	137			61
2,6-Dimethylphenol	203	49	140		79		159		
2,4-Dichlorophenol	210	45	140					125	96
2,4-Dibromophenol	239	40	153			184		120	98
2,4-Diiodophenol		72	166						98
2,4,6-Trichlorophenol		69	182			106			75
2,4,6-Tribromophenol		94	200	164		153	174	113	81
2,4,6-Triiodophenol		159	224				181		
2,4-Dinitrophenol		113		209		139		121	132
Picric acid		122				143			
1-Naphthol	279	94	192	144	105	143	217	88	56
2-Naphthol	285	123	154	110	84	169	210	125	107
o-Hydroxydiphenyl	275	58						65	76
p-Hydroxydiphenyl	306	165						177	151

6. TABLES OF PHYSICAL PROPERTIES

ALIPHATIC ALDEHYDES

	B.P.	2,4-Dinitrophenylhydrazone	p-Nitrophenylhydrazone	Semicarbazone	Dimethone	Dimethone anhydride	Oxime
Formaldehyde	(Gas)	166	182		189	171	
Acetaldehyde	20	168	129	169	141	174	
Propionaldehyde	49	155	124	89	155	143	
n-Butyraldehyde	75	123	91	104	134	141	
iso-Butyraldehyde	64	187	131	126	154	144	
n-Valeraldehyde	104	108			105		52
iso-Valeraldehyde	92	123	110	132	155	173	
Pivalaldehyde	75	209	119	190			41
n-Hexaldehyde	128	107	80	109	109		51
Diethylacetaldehyde	117	130		99	102		
n-Heptaldehyde	154	108	73	109	103	112	
n-Octaldehyde	170	106	80	98	90	101	60
n-Nonaldehyde	185	100		100	86		64
n-Decaldehyde	208	104		102	92		69
Acrolein	52	165	151	171	192	163	
Crotonaldehyde	102	190	185	201	194	167	119
Citral	229	110		164			
Chloral[†]	98	131					

[†] Hydrate b.p. 96, m.p. 53

ALIPHATIC KETONES

	B.P.	2,4-Dinitrophenylhydrazone	p-Nitrophenylhydrazone	Semicarbazone	Phenylhydrazone	Oxime
Acetone	56	128	149	190	42	59
Diethyl ketone	102	156	144	139		
Di-n-propyl ketone	144	75		133		
Diisopropyl ketone	124	88		160		
Di-n-butyl ketone	185			90		
Methyl ethyl ketone	80	111	129	146		
Methyl-n-propyl ketone	102	144	117	106		
Methyl isopropyl ketone	94	120	109	114		
Methyl-n-butyl ketone	128	107	88	125		
Methyl isobutyl ketone	117	95	79	132		
cycloPentanone	131	146	154	210	55	57
cycloHexanone	156	162	147	167	81	91
cycloHeptanone	180	148		162		
2-Methylcyclopentanone	139			184		
2-Methylcyclohexanone	165	137	132	197		43
3-Methylcyclohexanone	170	155	119	191	94	
4-Methylcyclohexanone	171	134	128	200	110	39
Camphor	209 (m.p.179)	177	217	238	233	119

6. TABLES OF PHYSICAL PROPERTIES

AROMATIC ALDEHYDES

	B.P.	M.P.	2,4-Dinitrophenylhydrazone	p-Nitrophenylhydrazone	Semicarbazone	Dimethone	Dimethone anhydride	Phenylhydrazone	Oxime
Benzaldehyde	179		237	192	224	195	200	158	35
o-Tolualdehyde	197		194	222	224	167	215	106	49
m-Tolualdehyde	199		194	157	216	172	206	91	60
p-Tolualdehyde	204		233	201	234			112	80
o-Chlorobenzaldehyde	213		206	249	226	205	225	86	76
m-Chlorobenzaldehyde	213		255	216	229			134	71
p-Chlorobenzaldehyde	214	47	265	237	232			127	110
m-Bromobenzaldehyde	234		257	220	205			141	72
p-Bromobenzaldehyde		67	257	208	228			113	
o-Nitrobenzaldehyde		44	265	263	256			156	103
m-Nitrobenzaldehyde		58	292	247	246	198	172	121	122
p-Nitrobenzaldehyde		106	320	249	221	190	222	155	133
Salicylaldehyde	197		252	228	231	211	208	143	63
p-Hydroxybenzaldehyde		116	280	266	224	189	246	177	72
Anisaldehyde	248		254	161	209	145	243	121	132
Phenylacetaldehyde	195	34	121	151	156	165		63	99
Cinnamaldehyde	252		255	195	215	219	175	168	139
Vanillin		81	269	226	239	197	228	105	117

216

AROMATIC KETONES

	B.P.	M.P.	2,4-Dinitrophenylhydrazone	p-Nitrophenylhydrazone	Semicarbazone	Phenylhydrazone	Oxime
Acetophenone	202	20	249	185	200	106	60
Propiophenone	218	20	191		182		54
Butyrophenone	229		190		188		50
o-Methylacetophenone	214		159		212		61
m-Methylacetophenone	220		207		205		55
p-Methylacetophenone	226	28	248	198	208	96	88
Benzophenone	306	49	238	155	167	137	144
Methyl benzyl ketone	216	27	156	145	190	87	69
1-Tetralone	129/12 mm		257	231	226	84	
Methyl 1-Naphthyl ketone		34			229	146	139
Methyl 2-Naphthyl ketone		56	262		236	177	145
m-Nitroacetophenone		81	228		257	135	132
o-Chloroacetophenone	229				160		113
m-Chloroacetophenone	228			176	232		88
p-Chloroacetophenone	236	20	231	239	203	114	94
o-Bromoacetophenone	112/10 mm				177		
m-Bromoacetophenone	131/16 mm				238		
p-Bromoacetophenone	256	51	230	248	208	126	129

217

6. TABLES OF PHYSICAL PROPERTIES

ALIPHATIC ESTERS

	B.P.	$d_{4°}^{20°}$	$n_D^{20°}$
Methyl formate	32	0.974	1.344
Ethyl formate	53	0.923	1.360
n-Propyl formate	81	0.904	1.377
iso-Propyl formate	71	0.873	
n-Butyl formate	106	0.892	1.389
iso-Butyl formate	98	0.876	1.386
sec -Butyl formate	97	0.884	1.384
tert -Butyl formate	83		
Methyl acetate	56	0.939	1.362
Ethyl acetate	77	0.901	1.372
n-Propyl acetate	101	0.887	1.384
iso-Propyl acetate	88	0.872	1.377
n-Butyl acetate	124	0.881	1.394
iso-Butyl acetate	116	0.871	1.390
sec -Butyl acetate	112	0.872	1.389
tert -Butyl acetate	97	0.867	1.386
n-Amyl acetate	148	0.875	1.402
iso-Amyl acetate	141	0.872	1.400

The $d_{4°}^{20°}$ is the density of the substance at 20°C relative to the density of water at 4°C. The $n_D^{20°}$ is the refractive index at 20° for the sodium D line (5893Å).

6. TABLES OF PHYSICAL PROPERTIES

ALIPHATIC ESTERS
(Continued)

	B.P.	$d_{4°}^{20°}$	$n_D^{20°}$
Methyl propionate	79	0.915	1.377
Ethyl propionate	98	0.892	1.384
n-Propyl propionate	122	0.882	1.393
iso-Propyl propionate	111		
n-Butyl propionate	145	0.875	1.401
Methyl n-butyrate	102	0.898	1.387
Ethyl n-butyrate	120	0.879	1.392
n-Propyl n-butyrate	142	0.872	1.400
iso-Propyl n-butyrate	128		
n-Butyl n-butyrate	165	0.869	1.406
Methyl iso-butyrate	91	0.888	1.383
Ethyl iso-butyrate	110	0.869	1.387
Methyl n-valerate	127	0.890	1.397
Ethyl n-valerate	144	0.874	1.400
Methyl iso-valerate	116	0.881	1.393
Ethyl iso-valerate	133	0.865	1.396
Methyl n-caproate	149	0.885	1.405
Ethyl n-caproate	165	0.871	1.407
Methyl stearate	M.p. 39		
Ethyl stearate	M.p. 33		

6. TABLES OF PHYSICAL PROPERTIES

<u>ALIPHATIC ESTERS</u>
(Continued)

	B.P.	$d_{4°}^{20°}$	$n_D^{20°}$
Methyl chloroacetate	129	1.234	1.422
Ethyl chloroacetate	142	1.150	1.422
Methyl crotonate	119	0.946	1.425
Ethyl crotonate	137	0.918	1.425
Methyl oxalate	M.P.54		
Ethyl oxalate	183	1.079	1.410
Methyl malonate	179	1.119	1.420
Ethyl malonate	197	1.055	1.414
Methyl succinate	195	1.120	1.420
Ethyl succinate	218	1.042	1.420
Methyl glutarate	109 /21	1.087	1.424
Ethyl glutarate	118 /15	1.023	1.424
Methyl adipate	121 /17	1.063	1.428
Ethyl adipate	134 /17	1.009	1.428
Methyl maleate	201	1.150	1.442
Ethyl maleate	220	1.066	1.440
Methyl fumarate	193	M.p.102	
Ethyl fumarate	214	1.052	1.441

6. TABLES OF PHYSICAL PROPERTIES

AROMATIC ESTERS

	B.P.	M.P.	$d_{4°}^{20°}$	$n_D^{20°}$
Methyl benzoate	199		1.089	1.517
Ethyl benzoate	212		1.047	1.505
n-Propyl benzoate	225		1.023	1.500
iso-Propyl benzoate	218		1.011	
n-Butyl benzoate	248		1.005	1.497
iso-Butyl benzoate	242		0.999	
Methyl phenylacetate	215		1.068	1.507
Ethyl phenylacetate	228		1.033	1.497
Methyl o-toluate	208		1.068	
Ethyl o-toluate	227		1.034	1.508
Methyl m-toluate	215		1.061	
Ethyl m-toluate	231		1.028	1.506
Methyl p-toluate	217	34		
Ethyl p-toluate	228		1.025	1.507
Methyl cinnamate	261	36		
Ethyl cinnamate	273		1.049	1.560
Methyl salicylate	223		1.184	1.537
Ethyl salicylate	231		1.125	1.522
n-Butyl salicylate	260			
Methyl m-hydroxybenzoate		70		
Ethyl m-hydroxybenzoate	295	73		
Methyl p-hydroxybenzoate		131		
Ethyl p-hydroxybenzoate	297	116		

6. TABLES OF PHYSICAL PROPERTIES

<u>AROMATIC ESTERS</u>

(Continued)

	B.P.	M.P.	$d_{4°}^{20°}$	$n_D^{20°}$
Methyl o-methoxybenzoate	245		1.156	1.534
Ethyl o-methoxybenzoate	248			
Methyl m-methoxybenzoate	237		1.131	1.522
Ethyl m-methoxybenzoate	251		1.100	1.515
Methyl anisate		49		
Ethyl anisate	261	7	1.103	1.524
Methyl o-chlorobenzoate	234			
Ethyl o-chlorobenzoate	243			
Methyl m-chlorobenzoate	231			
Ethyl m-chlorobenzoate	245			
Methyl p-chlorobenzoate		44		
Ethyl p-chlorobenzoate	238			
Methyl o-bromobenzoate	246			
Ethyl o-bromobenzoate	255			
Methyl m-bromobenzoate		32		
Ethyl m-bromobenzoate	259			
Methyl p-bromobenzoate		81		
Ethyl p-bromobenzoate	263			
Methyl o-iodobenzoate	278			
Ethyl o-iodobenzoate	275			
Methyl m-iodobenzoate	277	54		
Ethyl m-iodobenzoate	150 /15			
Methyl p-iodobenzoate		114		
Ethyl p-iodobenzoate	153 /14			

ALIPHATIC ALCOHOLS

	B.P.	$n_D^{20°}$	3,5-Dinitrobenzoate	p-Nitrobenzoate	3-Nitrophthalate
Methanol	65	1.3312	109	96	153
Ethanol	78	1.3624	94	57	157
1-Propanol	97	1.3854	75	35	145
2-Propanol	82	1.3776	122	110	153
1-Butanol	118	1.3993	64	36	147
2-Butanol	100	1.397	76	26	131
2-Methyl-1-propanol	108	1.3968	88	68	179
2-Methyl-2-propanol	83[a]	1.3878	142	116	
1-Pentanol	138	1.4100	46		136
2-Pentanol	119	1.4053	62		103
3-Pentanol	116	1.4077	100		121
2-Methyl-1-butanol	129	1.4087	70		158
3-Methyl-1-butanol	132	1.4084	62	21	166
2-Methyl-2-butanol	102	1.4052	118	85	
3-Methyl-2-butanol	113	1.3973	76		
2,2-Dimethyl propanol	113[b]				
cycloPentanol	141	1.4153	115	62	
cycloHexanol	161[c]	1.4656	113	50	160
Ethylene glycol	197	1.4274	169	141	
Glycerol	290	1.4729		188	

(a) m.p. 25   (b) m.p. 52   (c) m.p. 25

6. TABLES OF PHYSICAL PROPERTIES

AROMATIC ALCOHOLS

	B.P.	M.P.	3,5-Dinitrobenzoate	p-Nitrobenzoate	3-Nitrophthalate
Benzyl alcohol	205		113	86	183
2-Phenylethanol	220		108	63	123
1-Phenylethanol	203	20	94	43	
Cinnamyl alcohol	257	33	121	78	
Diphenyl carbinol	298	69	142	46	
Triphenyl carbinol		165			
o-Nitrobenzyl alcohol	270	74			
m-Nitrobenzyl alcohol		27			
p-Nitrobenzyl alcohol	185/12	93			
o-Aminobenzyl alcohol		82			
m-Aminobenzyl alcohol		97			
p-Aminobenzyl alcohol		65			
o-Chlorobenzyl alcohol	230	74	94		
m-Chlorobenzyl alcohol	234				
p-Chlorobenzyl alcohol	235	75			
o-Bromobenzyl alcohol		80			
m-Bromobenzyl alcohol	254				
p-Bromobenzyl alcohol		77			
o-Hydroxybenzyl alcohol		87			
m-Hydroxybenzyl alcohol		73			
p-Hydroxybenzyl alcohol		125			

6. TABLES OF PHYSICAL PROPERTIES

ALIPHATIC ETHERS

Ether	B.P.	$d_{4°}^{20°}$	$n_D^{20°}$
Dimethyl	(gas)		
Diethyl	34	0.714	1.353
Di-n-propyl	90	0.749	1.381
Di-iso-propyl	68	0.726	1.368
Di-n-butyl	141	0.770	1.399
Di-n-amyl	185	0.785	1.412
Di-iso-amyl	171	0.778	1.409
Di-n-hexyl	223	0.793	1.420
Di-n-heptyl	258	0.801	1.427
Di-n-octyl	288	0.806	1.433
Di-n-decyl	185/5	0.815	1.441
Methyl n-butyl	70	0.774	1.374
Ethyl n-butyl	92	0.749	1.382
Methyl n-amyl	99	0.761	1.387
Ethyl n-amyl	118	0.762	1.393
Methyl n-hexyl	126	0.772	1.397
Ethyl n-hexyl	142	0.772	1.401
Methyl cyclopentyl	105	0.862	1.420
Ethyl cyclopentyl	122	0.853	1.423
Methyl cyclohexyl	134	0.875	1.435
Ethyl cyclohexyl	149	0.864	1.435

6. TABLES OF PHYSICAL PROPERTIES

Ether	B.P.	$d_{4°}^{20°}$	$n_D^{20°}$
Ethyleneglycol dimethyl	83	0.863	1.374
Ethyleneglycol diethyl	123	0.848	
Diethyleneglycol diethyl	187	0.906	1.411
Benzyl methyl	171	0.965	1.501
Benzyl ethyl	186	0.948	1.496
Dibenzyl	299	1.042	
Tetrahydrofuran	65	0.889	1.407
Dioxan	101	1.034	1.417
Chloromethyl	59	1.015	
1-Chloroethyl	97	0.966	1.405
2-Chloroethyl	107	0.989	1.411
2,2'-Dichloroethyl	178	1.220	1.457

PHENOL ETHERS (AROMATIC ETHERS)

	B.P.	M.P.	$d_{4°}^{20°}$	$n_D^{20°}$	Nitro cpd.	Bromo cpd.	Sulfonamide	Picrate	Acid[†]
Anisole	153		0.996	1.518	94b	61f	113	81	
Phenetole	170		0.970	1.509	58c		150	92	
1-Phenoxypropane	188		0.949	1.501					
1-Phenoxybutane	208		0.934	1.497				112	
o-Methoxytoluene	171		0.980	1.517	69d	64h	137	119	101
m-Methoxytoluene	177		0.971		54e		130	114	110
p-Methoxytoluene	175		0.970		122b		182	89	184
o-Ethoxytoluene	184		0.952	1.504				118	
m-Ethoxytoluene	192							115	137
p-Ethoxytoluene	189			1.507				111	196
o-Chloroanisole	195				95d		131		
m-Chloroanisole	194								
p-Chloroanisole	198				98d		151		
o-Bromoanisole	210				106d		140		
m-Bromoanisole	210								
p-Bromoanisole	215				88d		148		
1-Methoxynaphthalene	265					46h	157	129	
2-Methoxynaphthalene	274	72				63h	151	117	
1,4-Dimethoxybenzene	212	56			72e	142i	148	48	

[†]Alkoxybenzoic acid from side-chain oxidation

6. TABLES OF PHYSICAL PROPERTIES

ALKANES AND CYCLOALKANES

	B.P.	$d_{4°}^{20°}$	$n_D^{20°}$
n-Pentane[†]	36	0.627	1.358
2-Methylbutane	28	0.620	1.354
n-Hexane	69	0.659	1.374
2-Methylpentane	60	0.653	1.372
n-Heptane	98	0.683	1.388
n-Octane	125	0.703	1.397
2,2,4-Trimethylpentane	99	0.688	1.389
n-Nonane	151	0.717	1.405
n-Decane	173	0.730	1.412
n-Undecane	196	0.740	1.417
n-Dodecane	216	0.750	1.422
cycloPentane	49	0.745	1.406
cycloHexane	81	0.779	1.426
cycloHeptane	118	0.811	1.445
Methylcyclohexane	101	0.769	1.423
Ethylcyclohexane	130	0.784	1.432
n-Propylcyclohexane	155	0.790	1.436
iso-Propylcyclohexane	155	0.802	1.441
trans-Decahydronaphthalene (Decalin)	185	0.870	1.470
cis-Decahydronaphthalene (Decalin)	193	0.895	1.481
1,2,3,4-Tetrahydronaphthalene (Tetralin)	207	0.971	1.543

[†]Methane, ethane, propane and the butanes are gases

ALKENES, CYCLOALKENES AND ALKYNES

	B.P.	$d_{4°}^{20°}$	$n_D^{20°}$
1-Pentene	30	0.641	1.371
2-Pentene	36	0.651	1.380
2-Methyl-1-butene	31	0.650	1.378
Trimethylethylene	38	0.662	1.388
Isoprene	34	0.681	1.419
1-Hexene	64	0.674	1.388
1-Heptene	93	0.697	1.400
1-Octene	121	0.716	1.409
1-Decene	169	0.742	1.422
1-Dodecene	80 /5	0.760	1.430
cycloPentene	45	0.772	1.420
cycloHexene	83	0.810	1.445
1,3-cycloPentadiene	41	0.803	1.443
1,3-cycloHexadiene	81	0.841	1.474
1-Pentyne (n-Propylacetylene)	39	0.695	1.385
2-Pentyne (Ethylmethylacetylene)	56	0.712	1.404
1-Hexyne (n-Butylacetylene)	71	0.717	1.399
1-Heptyne (n-Amylacetylene)	98	0.734	1.409
1-Octyne (n-Hexylacetylene)	126	0.748 (25°)	1.423 (25°)
1-Nonyne (n-Heptylacetylene)	151	0.760	1.423

AROMATIC HYDROCARBONS

	B.P.	M.P.	$d_{4°}^{20°}$	$n_D^{20°}$	Sulfonamide	o-Aroylbenzoic acid	Nitro cpd.	Bromo cpd.	Picrate
Benzene	80	6	0.879	1.501	153	128	89a	89g	
Toluene	110		0.867	1.497	137	138	71a		
Ethylbenzene	135		0.868	1.496	109	122	37a		
o-Xylene	144		0.880	1.505	147	129		88g	
m-Xylene	139		0.864	1.497	137	126	182a	72g	
p-Xylene	138	13	0.861	1.496	144	132	139a	75g	
n-Propylbenzene	159		0.864	1.493		126			
i-Propylbenzene	151		0.862	1.491		134			
Mesitylene	164		0.865	1.499		212	235b	224g	97
n-Butylbenzene	182		0.860	1.490		98			
t-Butylbenzene	169		0.867	1.493			62a		
Styrene	146		0.909	1.546				73i	
Napthalene	218	80				173	61c		150
Anthracene	340	216					226h		138
Phenanthrene	340	100					63g		143
Diphenyl	254	70				225	234d	164f	
Diphenylmethane	262	25					172a	64g	
Triphenylmethane	358	92					206c		

ALKYL HALIDES

Alkyl(a)	Chloride			Bromide			Iodide			S-Alkylisothiouronium picrate
	B.P.	$d_{4°}^{20°}$	$n_D^{20°}$	B.P.	$d_{4°}^{20°}$	$n_D^{20°}$	B.P.	$d_{4°}^{20°}$	$n_D^{20°}$	
Methyl							42	2.282	1.532	224
Ethyl	12			38	1.460	1.425	73	1.940	1.514	188
1-Propyl	46	0.889	1.388	71	1.435	1.355	102	1.743	1.505	177
2-Propyl	35	0.863	1.378	59	1.425	1.314	89	1.703	1.499	196
Allyl	45	0.940	1.416	70	1.432	1.470	100	1.777	1.578	155
1-Butyl	77	0.886	1.402	101	1.274	1.440	129	1.616	1.499	177
2-Butyl	68	0.874	1.397	91	1.256	1.437	118	1.592	1.499	166
2-Methyl-1-propyl(b)	69	0.881	1.398	91	1.253	1.435	119	1.602	1.496	167
2-Methyl-2-propyl(c)	50	0.846	1.386							
1-Pentyl	105	0.882	1.412	129	1.219	1.445	155	1.512	1.496	154
2-Pentyl	97	0.873	1.408	117	1.212	1.442	142	1.510	1.496	
3-Pentyl	96	0.872	1.408	118	1.211	1.443	142	1.511	1.497	
3-Methyl-1-butyl(d)	99	0.872	1.409	119	1.213	1.442	147	1.503	1.493	173
2-Methyl-2-butyl(e)	85	0.865	1.405				128	1.479		
2,2-Dimethyl-propyl(f)	85	0.879		109	1.225					

(a) Several names transgress IUPAC rules. Thus 2-propyl chloride should be 2-chloropropane. We feel that the unambiguity, simplicity and correlation with the names of alcohols justify the transgression.

Common names (b) iso-Butyl, (c) tert-Butyl, (d) iso-Amyl, (e) tert-Amyl, (f) neopentyl.

6. TABLES OF PHYSICAL PROPERTIES

ALKYL HALIDES
(Continued)

Alkyl	Chloride			Bromide			Iodide			S-Alkylisothio-uronium picrate
	B.P.	$d_{4°}^{20°}$	$n_D^{20°}$	B.P.	$d_{4°}^{20°}$	$n_D^{20°}$	B.P.	$d_{4°}^{20°}$	$n_D^{20°}$	
1-Hexyl	134	0.878	1.420	154	1.175	1.448	180	1.437	1.493	157
1-Heptyl	159	0.877	1.426	178	1.140	1.451	201	1.373	1.490	142
1-Octyl	182	0.875	1.431	200	1.112	1.453	221	1.330	1.489	134
cycloPentyl	114	1.005	1.451	137	1.387	1.489		1.709	1.547	
cycloHexyl	142	0.989	1.462	164	1.336	1.495		1.624	1.547	
Benzyl	177	1.100	1.539	198	1.438					187

ARYL HALIDES

	B.P.	M.P.	$d^{20°}_{4°}$	$n^{20°}_D$	Sulfonamide	Nitro cpd.	Acid[†]
Chlorobenzene	132		1.107	1.525	143	52a	
Bromobenzene	156		1.494	1.560	160	75a	
Iodobenzene	188		1.831	1.620		174b	
o-Chlorotoluene	159		1.082	1.524	126	45a	141
m-Chlorotoluene	162		1.072	1.521	185	91a	158
p-Chlorotoluene	162		1.071	1.521	143	38b	242
o-Bromotoluene	181		1.425		146	82a	148
m-Bromotoluene	184		1.410		168	103b	155
p-Bromotoluene	185	28	1.390		165	47c	251
o-Iodotoluene	207		1.698			103b	162
m-Iodotoluene	204		1.698			108c	186
p-Iodotoluene	211	35					269
o-Dichlorobenzene	180		1.305	1.552	135	110a	
p-Dichlorobenzene	174	53			180	54c	
o-Dibromobenzene	224		1.956	1.609	176	114a	
p-Dibromobenzene	219	89			195	84c	
p-Bromochlorobenzene	195	67				72	
1-Chloronaphthalene	263		1.191	1.633	186	180c	
2-Chloronaphthalene	265	61			232	175b	
1-Bromonaphthalene	281		1.484	1.658	193	85	
2-Bromonaphthalene	282	59			208		
1,2-Dichloronaphthalene	296	35					

[†]Halogeno acid from side-chain oxidation.

INDEX

INDEX

INDEX

INDEX

DATE DUE